T0184513

Modeling Dynamic Systems

Series Editors

Matthias Ruth
Bruce Hannon

Springer Science+Business Media, LLC

MODELING DYNAMIC SYSTEMS

Modeling Dynamic Biological Systems
Bruce Hannon and Matthias Ruth

Modeling Dynamic Economic Systems
Matthias Ruth and Bruce Hannon

Dynamic Modeling in the Health Sciences
James L. Hargrove

———————————————————————————————

Forthcoming:

Computer Modeling and Simulation in Pre-College Science Education
Nancy Roberts, Wallace Feurzeig and Beverly Hunter

James L. Hargrove

Dynamic Modeling
in the Health Sciences

With 122 Illustrations and a CD-ROM

Springer

James L. Hargrove
Department of Foods and Nutrition
University of Georgia
Athens, GA 30602
USA

Series Editors:
Matthias Ruth
Center for Energy and
   Environmental Studies
   and the Department of Geography
Boston University
675 Commonwealth Avenue
Boston, MA 02215
USA

Bruce Hannon
Department of Geography
220 Davenport Hall, MC 150
University of Illinois
Urbana, IL 61801
USA

Library of Congress Cataloging-in-Publication Data
Hargrove, James L.
   Dynamic modeling in the health sciences / James L. Hargrove.
      p.   cm. — (Modeling dynamic systems)
   Includes index.

   ISBN 978-1-4612-7228-1   ISBN 978-1-4612-1644-5 (eBook)
   DOI 10.1007/978-1-4612-1644-5

   1. Medicine—Computer simulation.   2. Medical sciences—Computer simulation.
   3. Dynamics—Computer simulation.   I. Title.   II. Series.
   R859.7.C66H37   1998
   610'.1'13—dc21   97-39337

Printed on acid-free paper.

Additional material to this book can be downloaded from http://extras.springer.com

Production coordinated by Diane Ratto, Michael Bass & Associates, and managed by Lesley
Poliner; manufacturing supervised by Thomas King.
Typeset by G&S Typesetters, Austin, TX.

9 8 7 6 5 4 3 2 1
SPIN 10570120

*This book is dedicated to
all mentors and students who take joy in discovery
and seek to understand the temporal relations of things.
Notwithstanding this sentiment,
it is for Diane, Katharine, and John,
who had the good sense to go fishing in the summer of 1997
and let Dad write!*

Series Preface

The world consists of many complex systems, ranging from our own bodies to ecosystems to economic systems. Despite their diversity, complex systems have many structural and functional features in common that can be effectively simulated using powerful, user-friendly software. As a result, virtually anyone can explore the nature of complex systems and their dynamical behavior under a range of assumptions and conditions. This ability to model dynamic systems is already having a powerful influence on teaching and studying complexity.

The books is this series will promote this revolution in "systems thinking" by integrating skills of numeracy and techniques of dynamic modeling into a variety of disciplines. The unifying theme across the series will be the power and simplicity of the model-building process, and all books are designed to engage the reader in developing their own models for exploration of the dynamics of systems that are of interest to them.

Modeling Dynamic Systems does not endorse any particular modeling paradigm or software. Rather, the volumes in the series will emphasize simplicity of learning, expressive power, and the speed of execution as priorities that will facilitate deeper system understanding.

Matthias Ruth and Bruce Hannon

Preface

The chemical reactions must be balanced so delicately that, through regeneration, the body components remain constant in total amount and
in structure. This constancy is not to be taken as an indication that the
structural matter of the living organism is inactive and takes little part in
metabolism.

Rudolf Schoenheimer, M.D., *The Dynamic State of Body Constituents,*
Cambridge, Mass.: Harvard University Press, 1942.

New Tools for Quantitative Thinking

Rudolf Schoenheimer was among the first scientists to discover proof that the
human or animal body was subject to renewal and regeneration. He and his
colleagues did this by using isotopic tracers to show that a portion of the fat,
protein, and carbohydrate in the tissues is replaced daily. In a sense, this
marked the beginning of the ability to study rates of change, or *The Dynamic
State of Body Constituents*. The new techniques also vastly augmented the
ability to measure quantities of substances in the body. Until the advent of the
personal computer and software for dynamic modeling, however, it was
difficult and tedious to make use of this wealth of quantitative information
about dynamic systems.

This book is intended for those who have not previously used a personal
computer as a tool for quantitative thinking. It will focus on ideas that originated in classical studies in the health sciences, including energetics, nutrition, physiology, genetics, and epidemiology. Throughout the book, I use the
term "students" to indicate people of any age or level of training who have
not previously used software for dynamic modeling, for that is the intended
readership. The book is *not* directed toward the biophysicist, statistician, or
biomedical engineer who has advanced training in mathematics, and it presupposes no knowledge beyond basic algebra and the use of a computer,
mouse, and graphical user interface. This book is also not intended only to
teach how to use the **STELLA®** modeling program; in a broader sense, it is
about thinking about time and change.

STELLA®: A Finger Pointing Toward the Stars

Students who train for careers in health care or medical science will spend
their professional lives dealing with strategies to understand, preserve, and
foster good health, and the consequences of failure to maintain health. Like-

wise, the biomedical research scientist must identify problems in subjects such as aging, cancer, diabetes, and heart disease that are amenable to experimentation and that merit funding from a national agency. Undertaking a career in health care means that one must prepare to think about, and deal with, changes that occur over time. Some patients will have made wise choices and come close to realizing their full potential. Inevitably, there will be others whose health has deteriorated because of poor choices, difficult circumstances, family history, accidents, or the inexorable process of aging.

The true professional is challenged to do more than just deal with the consequences of disease. It is also essential to be able to interpret quantitative data in the light of progress in medical science. To know what is normal, or to diagnose a specific condition, one must be able to assess function through appropriate laboratory tests. To interpret this information, it is critical to read and understand the new scientific literature. If one now adds these key elements together, it becomes clear that there is a need to create a model that encompasses the dynamics of the human body. However, this should be more than a mental model; it must be a dynamic model that is capable of incorporating quantitative information and new discoveries. Indeed, a discipline capable of creating such models could be called the System Dynamics of Human Health.

The advent of the personal computer now enables any intelligent individual to begin making use of dynamic modeling. No longer must a student develop very sophisticated facility with mathematics, for new software enables one to create dynamic models with the only prerequisites being a moderate understanding of algebra, a willingness to learn to use the software, and a desire to think about human biology in terms of the dynamics of our physiological states. This book is intended to help the student develop facility with these tools to accomplish what was heretofore impossible: an ability to generate quantitative models based on the Biokinetic Database found in published research literature.

People learn by doing. This was true when you learned how to tie your shoes, to ride a bicycle, cook, play baseball, shoot baskets, or drive a car. Whereas most people remember nearly everything that they learn to do, they remember surprisingly little of what they may have read or viewed. To be effective learners, students must gather information, formulate models, and sit down alone or together at a computer, and begin to create dynamic models of the human body. It is impossible to do this without thinking. With the use of this software, it will be a kind of thinking that has never been possible before. Students will be able to make predictions about change over time. They may ask questions concerning strategies to deal with real human problems, for every one of us faces the consequences of change over time as we live, learn, and age. Though this approach may be hypothetical, using the technique of computer simulation, the problems are real: accumulation of cholesterol in the blood, prevention of obesity, the development of anemia and cancer, and whatever else your mind may lead you to consider.

This book demonstrates some simple dynamic models that have been created using **STELLA®** software for the personal computer. These are to be regarded as starting points, for in creating a dynamic model, one must simplify. By definition, then, the models are incomplete. Students will be aware that much more information is available for each of these examples in the form of scientific publications in library journals and reference texts. Just as soon as one has learned how to use simulation and modeling to grapple with biological problems, one should focus on a problem of personal interest. Successful students will seek out better information, develop hypotheses, and strive to create models to help test ideas. To learn how to use any tool, one must first pick it up, study it, and engage in play. Let us begin with this sense of playfulness.

Learning Objectives and Methods of Procedure

How can one prepare to become engaged in the process of science, when one may not have access to laboratory facilities? The traditional method of having laboratory demonstrations does not actively engage the student in the key components of science, and it has become rare for advanced classes to conduct laboratory sessions, as costs for equipment and materials have soared, and proscriptions against use of animals for demonstrations have become nearly insurmountable. An alternative approach is to combine extensive reading in the scientific and medical literature with group discussions and projects carried out with the computer. Given these tools, it is possible to think, to plan, and to experiment—at least by simulation. The computer generates dynamic behavior, and the theory of science becomes open to every person willing to learn how to operate software that can actually restore a sense of fun and adventure. You may not be able to fly the jumbo jet, but you will fly the flight simulator.

Each subject area in science has a unique language that contains specialized terms and jargon. This book is presented on a level that assumes the student is already familiar with basic concepts in biology and chemistry, and is pursuing a career in a biological or health science. In terms of the educational objectives or "Taxonomy of Learning" stated by Bloom (1956), students should be working toward the highest order of skills in which information is used to originate new ideas and ways are found to test these ideas. In this process, **STELLA®** provides a laboratory that truly permits one to ask: If this is how the system is designed, then how will it behave in the face of any change? Will change be adaptive and potentially beneficial, or will the system break down into unstable oscillations or demonstrate chaotic behavior?

It is expected that students will learn any unfamiliar scientific terms and be able to use them in the context of an organized model. Facts can be thought of as an organized set of terms used to appropriately describe or summarize a system of interest. Mentally, one develops a concept of the components of a system and how it operates. To do so, one applies a set of logical rules that can be stated quantitatively. The aim of this book is that the student be able to work through this hierarchy of learning to be able to think and predict possible outcomes of what would happen to a health system if one of its components were altered, resulting in disease.

There are many methods of approaching the problem of hypothesis testing, but the following procedure has been used in this text:

1. Determine if a general model for the special subject or disease has been described by research scientists, and summarize the findings in a brief manner. Expect to read some of the primary references, and use resources in the library and the Internet to update the information when time permits.
2. Ask if a kinetic model of the system of interest has already been published. When time is short, one does not need to "reinvent the wheel." A list of some published models is provided in the last chapter.
3. Determine if equations have been published that state quantitative relationships; these are excellent resources for model construction. Also look for published references in which the relationships have been tested experimentally, and in which the components of each system have been identified.
4. Translate the information into a kinetic model and test the effects of changing specific relationships. Look for important control points and feedback relationships. Print the outcome; document the approach taken, and save all files for preparing reports.
5. Test the model under a variety of conditions. Do the outcomes resemble patterns observed in the biomedical literature? In what ways could the model be modified, either to make it more inclusive, or to simplify if it has grown overly complex? Does it suggest control systems that may not have been studied in detail? If so, how could one go about this? Do you have ideas about why the control systems are needed, and how they might have developed or evolved?
6. Use your model to predict outcomes, and ask how you could use the theoretical results to plan experiments. What would you need to measure, and when would you plan to take your samples? In the area of applied medicine, does your model suggest any health consequences of various human behaviors? Could you use the data to help counsel a person who is interested in preventing disease, maintaining health, or dealing with a preexisting disorder?
7. Go back to the medical literature and seek studies that may have attempted to answer the questions raised by your model and its predictions. Is there reasonable agreement? Are future studies warranted?

Caveat

It is well said that one ought not to mistake a finger pointing at the moon for the moon itself. This book is about the software and the accompanying models; it is one person's interpretation of the information, and therefore limited.

The message that runs throughout the chapters is that **STELLA**® and similar software enable a new kind of thinking, a novel way of interpreting the scientific literature, and a new use for the personal computer. I am an advocate who believes the new capabilities are potentially revolutionary, and that the success of this effort depends on the truth of these premises. Try the software, run the models, and decide for yourself if you agree. Welcome to the Revolution!

James L. Hargrove

Reference

Bloom, B.S., ed. *Taxonomy of Educational Objectives. Handbook I: Cognitive Domain*. New York: Longman Press, 1956.

Acknowledgments

This book is dedicated to the mentors who have inspired this work, and to my students and colleagues. Timely instigation and inspiration came from Manuel Ramos, Paul Buck, Alan B. Bond, Eldon Gardner, LeGrande C. Ellis, Daryl K. Granner, and Edward J. Gallaher. Dr. Dora B. Goldstein kindly discussed her role in reviewing Cheston Berlin and Robert Schimke's published work on kinetic modeling of enzyme induction.

The models concerning specific disease states were largely developed by my students, as follows: Ms. Joi Foss, Cholesterol Metabolism; Ms. Kimberly Porter Smith, Bone Remodeling; Ms. Laurie Weissinger, Risk for Disease and Fat and Energy Balance; and Ms. Catherine Chung, Tumor Progression. The excerpts used here are meager in comparison to the extensive literature reviews that supported their model development; the full bibliographies are included in their Master's Theses. Their work was supported in part by a Higher Education Challenge Grant from the United States Department of Agriculture. I thank Hank Taylor and John Sterman of the Sloan School of Management at MIT for providing the **STELLA®** model of insulin dynamics that was modified for use in Chapter 24; Dr. Diane K. Hartle and Dr. Robert Gutman for reviewing the chapters, and Ms. Alison Brister and the staff in our Computer Laboratory for enabling this work to be used in the classroom.

James L. Hargrove

Contents

1

Discoveries with a Computer: Dynamic Behavior

> The important thing is not to stop questioning . . . . It is enough if one
> tries merely to comprehend a little of this mystery every day. Never lose
> a holy curiosity.
>
> Albert Einstein

In his book, *The Night Country*, Loren Eiseley recounted an archaeologist's experience of excavating an Egyptian pyramid as follows: "You crawl along some dark corridor on hands and knees, past falls of rock; the light of the lamp gleams on minute crystals in the stratified walls; beyond, the corridor disappears into the blackness. You turn corners, feeling your way with your hands; the workmen have been left behind, and suddenly, you realize you are alone in a place which has not heard a footfall for nearly fifty centuries."

Think of Charles Darwin on the *Beagle*, Alfred Russell Wallace in Sumatra, Gregor Mendel in an Austrian monastery, Albert Einstein working as a Swiss patent clerk, Barbara McClintock doing maize genetics at Cornell, Watson and Crick having a beer in a pub near the Cavendish Laboratories, or any of a hundred other scenarios from the history of science, and what comes to mind? Surely there is a pattern of discovery that is one of personal quest, one that is peculiar not to institutions, but rather to unique individuals working creatively in isolation or in small, intense groups, whether unfunded or underfunded, perhaps in keen competition for a major scientific prize.

Each published discovery began with an insight in a single human mind, and only incidentally does this occur in institutions designed to foster higher thinking. The human sense of curiosity and the need to explore and experience newness may have been codified in the process of scientific discovery, but it predated the scientific method, and one could argue as strongly con or pro that universities advance insight. Even if we will not be the first person to direct a lamplight through the window into King Tutankhamen's tomb, each of us will experience moments of revelation. And even though the weekly laboratory meeting may stimulate one to think, the moment of insight will always remain a singularity, something that occurs most often late at night in the mind of a person who has become obsessed with a particular problem, and who can not rest until a solution comes.

This chapter asks the reader to set up a computer model as a playful game or experiment with a serious goal. One is asked to gain insights about a system that shows erratic, indeed, chaotic behavior. One is asked to understand why the behavior occurs and if it can be controlled, and may incidentally learn something important about the way **STELLA**® operates—something inherent in the equations it uses, which are called finite difference equations. First, however, permit me to tell two anecdotes. One is a case in which I had an insight as a result of conducting a simple experiment in a laboratory; and the second (in Section 1.3) is a similar insight that occurred because I was finally using **STELLA**® to simulate what would happen if . . . but let me tell the story.

1.1 A Small Discovery in a Remote Place

My first experience that conveyed this sense of insight was very simple. I was working on a doctoral project intended to test the control of contractions of a certain kind of smooth, or visceral, muscle in response to lipid mediators called prostaglandins. I would tie the tissue preparation to a device that measured contractility, and immerse the tissue in a warm, oxygenated saline bath. At first, the preparation would lie inertly in the bath, but after a time, would begin to contract and produce vaguely sinusoidal waves on a chart recorder. I would add the compounds of interest, and after a time, remove the spent fluid and add fresh saline solution. Once again, the preparation would become inactive, and usually begin to contract again after fifteen minutes or so. Whereas some preparations were very active, others seemed balky and recalcitrant, never producing contractions.

One day, what should have been obvious struck me. This was an isolated tissue preparation with no nerves to initiate muscular contractions, and at any rate, visceral muscle tends to set its own rhythms with autonomous pacemakers. Could it be that the preparation was producing a mediator of some sort that caused the contractions to begin, and only began to respond after the mediator had accumulated? And that when I removed the old bathing medium, I was also removing the mediators, thereby ending the contractions?

So without running upstairs to ask my mentor for permission, I set up a preparation that began contracting, and then removed the medium and set it aside in a warm bath bubbling with oxygen. When I added fresh medium, the preparation lay there, motionless. In half an hour, it started contracting again, but stopped when I removed the medium. Then I added back the first medium, and the tissue responded as if it had just met an old friend—the contractions began immediately and vigorously! To prove the point, I extracted lipid-soluble materials from the medium and showed that the tissue was truly producing and responding to prostaglandins that were similar, if not identical, to the ones we had obtained from a pharmaceutical company.

The notion of performing this simple test would never have occurred to me had I not been seated in a small room with nothing to do but observe and think for hours at a time. It was not a great discovery, as these things go. I was not Einstein clerking in the Swiss patent office, thinking about time and space; I was not Barbara McClintock working out the genetics of transposons in an agricultural field station. But the hair did stand up on the back of my neck, I did tear the tracing off the chart recorder and run up three flights of stairs to show my mentor, and I did feel that I knew something that day that had been going on for millions of years, totally unsuspected. The fact that these substances may only have been important to male rabbits intent upon reproduction was immaterial. It was not so much the subject that was important, it was the impact it made on me as a developing scientist, shaping who I was to become.

My mentor later told me that the man who first observed contractility in the testicular capsule had set up the experiment in a physiology class as a way of demonstrating that nine out of ten experiments will fail. By succeeding, his experiment failed; he instead demonstrated that one out of ten scientific experiments will succeed and yield something new.

1.2 Discoveries with a Computer

I have occasionally had other moments of discovery while working in various laboratories, but was also vaguely familiar with the idea that some discoveries could occur by application of theory in advance of experimental testing. There is, for example, Albert Einstein's humorous account of one of the first tests of the Theory of Relativity. In reference to Max Planck, he said:

"He was one of the finest people I have ever known . . . but he really didn't understand physics, [because] during the night of the eclipse of 1919 he stayed up all night to see if it would confirm the bending of light by the gravitational field. If he had really understood [the general theory of relativity], he would have gone to bed the way I did."

Although the principal tools Albert Einstein used were a pencil and paper, these were only devices to record the workings of his mind using the true tool of mathematics. Often, he would attempt a problem for which he was not well prepared, and would find a colleague who could direct him to a different approach. Progress often necessitated learning new kinds of mathematics.

Other discoveries that involve pure theory are worth recounting, and some of the best were told by James Gleick in his best-selling book, *Chaos, Making a New Science*. Of these, the story that is most germane could be the work of Robert May on an example of what became the theory of chaos.

The reason the story is relevant is that Robert May was working with one of the best-known equations in population biology, the logistic difference equation, using nothing more than a hand-held calculator to determine outcomes. The equation is given as:

$$x_{next} = rx(1 - x) \qquad (1.1)$$

Dr. May was attempting to solve what happens "when a population's rate of growth, its tendency toward boom and bust, passed a critical point." (Gleick,

p. 70). And what Dr. May found was that the outcome depended on the value of r, giving stable periodic behavior for values less than 3, but that the period became very unstable, or chaotic, for the numbers between 3 and 4. Above a value of 4, the population became extinct. The outcome was very sensitive to initial conditions, and the papers describing the conclusions were important enough to be published in several preeminent scientific journals.

For Dr. May, and most readers of *Chaos*, the experience proved compelling. Even vicariously, the sense of discovery was enough to keep one up at night, turning pages, wondering what the next great finding would be. But perhaps most telling in the context of the present book was the statement on page 80:

"The world would be a better place, May argued, if every young student were given a pocket calculator and encouraged to play with the logistic difference equation. That simple calculation . . . could counter the distorted sense of the world's possibilities that comes from a standard scientific education." (p. 80)

In a moment, we shall do exactly that with an even better set of tools: your personal computer and **STELLA**® software.

1.3 From the Bench to the Mind and Back Again

Here follows the second story on insight I promised to tell.

I worked for several years with an enzyme called tyrosine aminotransferase, of which a graduate student at another university once wrote, "Tyrosine aminotransferase is like the proverbial Lady in Red; everyone wants to know her, but no one wants to admit it." What I thought was worth knowing about this "Lady," who caused me to sleep on a cot in a laboratory on more than one night, was that the enzyme and the messenger RNA on which it was encoded were both extremely unstable in living tissue. Unstable, at least, for mammalian enzymes, which are paragons of duration in comparison to their microbial counterparts.

In this capacity, I was very aware that several of this enzyme's "cousins," for instance alanine aminotransferase and aspartic aminotransferase, were quite stable by comparison. All were made in liver of the same amino acid building blocks, and all existed in the same part of the cell, insofar as anyone knew. So why should tyrosine aminotransferase and its mRNA both be unstable, when the other transaminases and their mRNAs were highly stable? It plainly had nothing whatsoever to do with the enzyme's role as a catalyst. What made the problem especially interesting was that the proteins and their mRNAs were made of different chemical substances—amino acids versus nucleotides—yet both kinds of macromolecule seemed to be inherently stable. Why were particular mRNAs and proteins unstable?

To help answer this question, I asked a colleague, Professor Frederick Schmidt, to write a computer program that would allow us to test the coordinate "induction" of a messenger RNA and the protein it encodes. Fred was happy to tackle a theoretical problem, but it required roughly six months for him to find time to solve the differential equations, program them in BASIC, and teach me how to use the software. Then, when I asked him to change some parameters, he explained that the problem became increasingly difficult.

So I used his original program to analyze the time course of enzyme induction, and concluded that there must be specific segments or domains in tyrosine aminotransferase and its messenger RNA that caused them to be unstable—that this could not have happened by chance. During that time, however, I experienced the feeling again of looking back through time perhaps a billion years, seeing that evolution must have selected this outcome for an extremely important purpose.

The purpose of creating unstable messenger RNAs and enzymes was to deal with change. It is physically impossible for the concentration of a stable enzyme or messenger RNA to change rapidly by usual mechanisms. In fact, it is such a difficult feat that an entirely new class of regulatory enzymes had to be invented, the most notable being the protein kinases and protein phosphatases. For whereas it is very difficult to change the concentration of a preexisting enzyme quickly, it is not difficult to cause it to be switched on and off.

This was the first great insight that I ever experienced using a theoretical tool. I understood that the RNA world had to speak to the protein world (and listen to it), even if molecular biologists who work with messenger RNA tended not to associate much with protein chemists. And one happy day at a scientific meeting, a colleague heard my story and asked, "Don't you know about **STELLA®**? You could have solved that programming problem yourself in ten minutes."

That work on the kinetics of gene expression caused me to understand that time could not be taken for granted, and that the greatest insights imaginable could come from studying what is familiar and close at hand. However, one needed a new tool, a quantitative one—like **STELLA®**. A tool that could help other people mine simple equations that lie near at hand, overlooked and unexplored. A tool that could help "outsiders" who may be working in isolation or in offices no bigger than monk's cells to generate insights that merit publication in the scientific literature.

A tool that does not require a federal grant or the imprimatur of the hierarchy.

1.4 The Chaos Game

The finite difference equation that Robert May was investigating predicts the growth of populations over time as a function of initial population density and reproductive rate. What he discovered was that under certain conditions, the equation begets instability. The equation is very simple to create using **STELLA®**, and instead of providing you with a model in this case, I will ask you to open the software and build the model yourself. No one learns to skate without taking a tumble or two, so please do this, it will only take a moment, and who knows what you may learn? Here are the steps to follow.

You will be using **STELLA**® and the icons on the tool bar to set up a diagram that looks like the one in Figure 1.1. First, open **STELLA**®. Select the stock icon, click the mouse button to put it in place, and type in a name (X is good). Select a flow icon, click and drag so it enters the stock icon and causes the stock icon to change color, then let go of the mouse button. Label the flow Delta X. With the mapping symbol showing in the left margin, double click on the flow icon, and select the Biflow option (because material will need to flow in and out through this single channel). Add a circular converter labeled R, and use the arrow (connector) to connect R to the converter on the flow. If your diagram looks like Figure 1.1, you are ready to enter the very simple equation.

Here are the equations as listed in the lower level of **STELLA**®:

```
X(t) = X(t - dt) + (Delta_X) * dt
INIT X = 0.10
Delta_X = R * X * (1 - X) - X
R = 2
```

To enter the algebra, open the stock labeled X (click twice quickly), and enter a starting value of 0.1. You will be asking **STELLA**® to tell you how this changes over time. Begin with a value equal to 2 in the converter labeled R; you will change it in a moment. Open the flow icon, and enter the values, $R * X * (1 - X) - X$. **STELLA**® will prompt you to do this by displaying R and X as two necessary input variables (see the dialog box in Figure 1.2). Make sure the final X in the equation is not within the parentheses; the value of X must be *subtracted* from the product of $(R * X * (1 - X))$ after each iteration. (Stocks normally accumulate everything that enters them through an *uniflow*

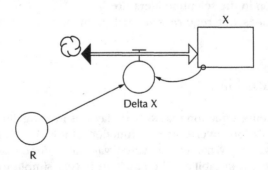

FIGURE 1.1

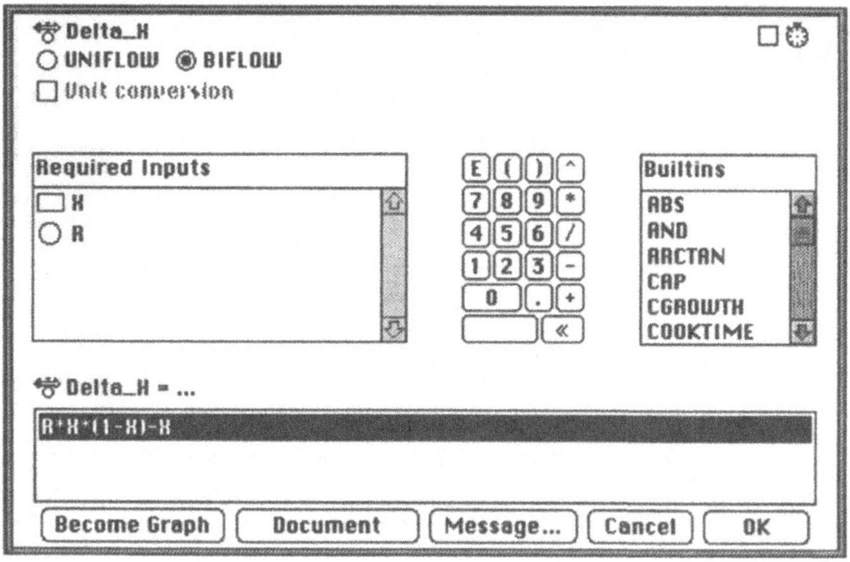

FIGURE 1.2

during an interval; by choosing a *biflow*, we are allowing the stock to subtract the prior value of X at each iteration.) Finally, begin by opening Time
Specs from the Run menu and choosing a time interval of 100 or more, and a
value for the calculation interval, DT, of 1.0. You will change this in a moment
just to see what happens. Now you should be ready to run your model, but
you can experiment by changing values of R and X and DT, and just play to
see what happens next and find out if any patterns emerge.

Note that this equation simulates population growth, but is scaled to a
peak value of 1.0. If you would prefer to imagine that you are running a
colony of rabbits, for instance, you can multiply X times a number, say 1000,
and watch your bunnies multiply as Easter approaches!

Now open a graph and select X as the variable to be plotted. It may be
worthwhile to print the graphs that you obtain in this series, so you will have
a record of your experiments. Figure 1.3 is a graph of an outcome for R = 2.

Now try R = 3.6 or other numbers between 3 and 3.9. Figure 1.4 gives an
example.

Compare that to the graph for R = 3.9 in Figure 1.5.

Just for fun (or perhaps for total consternation) try setting the value of
R to any number greater than 4 and see what happens to your bunnies. Was
that a wise decision as master of the colony?

Now it is time to start thinking. First you need to know that the mysterious
parameter R is *an intrinsic rate of increase*; it is how fast your bunnies are
multiplying within the confines of your colony. Comparing the outcomes for
R = 2 and above, it is evident that what happens between values of 2 and 3
is fairly humdrum. You start with a number, it grows, and then stops grow-

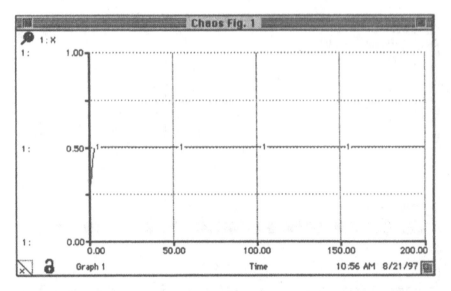

FIGURE 1.3

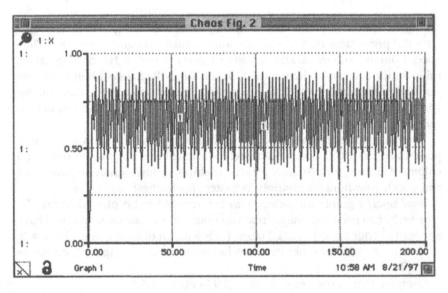

FIGURE 1.4

ing. X has reached a maximum from which it never varies or varies rather little. However, what happens when values of R are above 3? The population begins to oscillate, but in a rather stable manner at first. The closer the number gets to 4, the more erratic the oscillations become. If you want to confirm that, choose a smaller value for your time specs—40 or 50, perhaps. You will

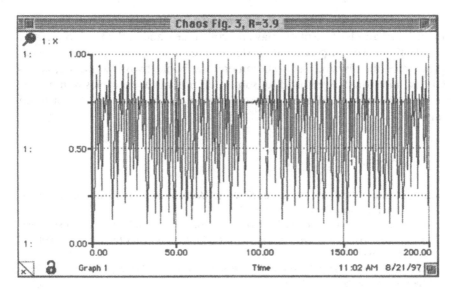

FIGURE 1.5

observe that the population swings predictably at low numbers and wildly at high numbers.

This simple model allows considerable insight into the behavior of some wild populations of animals, but there are many cases in which chaos theory is thought to affect medical problems; ventricular fibrillation is one possible example. But we are not done learning about this problem yet.

A finite difference equation is characterized by a fixed calculation interval. The equation you have just solved has a continuous counterpart, an analytical version for which a solution exists at every moment in time. So find out what happens if you change DT, which you set at 1.0 for the previous examples. Compare the outcomes as you reduce DT to 0.5, 0.25, and 0.1. Make it as small as you want, as long as it is greater than zero. What happens to the outcome for a particular value of R?

I do not wish to tell you everything about this equation. Just build confidence that what you are observing is a little slice of reality that you are trying to understand. You can find out what happens for any combination of the parameters R, X, or DT. However, ask yourself this question: If you were working with a more complex model, should you take the chance that you were accidentally introducing wild oscillations because you had chosen a large value of DT? Probably not; nature is dicey enough without our assistance. Models that contain some values that change rapidly and others that change slowly are called *stiff models* in the modeling world. In general, keep your calculation intervals short, or at least determine what happens if you decrease them. The calculation interval should not exceed one-half the shortest time constant in your model. So if one of your parameters has a rate constant of 1.0, your value of DT should not exceed 0.5.

More extensive discussions of the logistic growth equation can be found in the original articles by Robert May and colleagues, and in the book by Brown and Rothery. For inspirational stories concerning people who have made great scientific discoveries, I recommend the books by Barbara Cline, James Gleick, and Horace Freedland Judson. And for comical examples of how systems in the real world malfunction, *Systemantics* by John Gall is not to be missed!

References

Brown, D., and P. Rothery. *Models in Biology: Mathematics, Statistics, and Computing*. New York: John Wiley and Sons, 1993.

Calaprice, Alice. *The Quotable Einstein*. Princeton: Princeton University Press, 1996.

Cline, Barbara Lovett. *Men Who Made a New Physics*. 2nd ed. Chicago: University of Chicago Press, 1987.

Eiseley, Loren. *The Night Country, Reflections of a Bone-Hunting Man*. New York: Charles Scribner's Sons, 1971.

Gall, John. *Systemantics, The Underground Text of Systems Lore (How Systems Really Work and How They Fail)*. Ann Arbor, Mich.: The General Systemantics Press, 1988.

Gleick, James. *Chaos, Making a New Science*. New York: Penguin Books, 1987.

Judson, Horace Freeland. *The Eighth Day of Creation, Makers of the Revolution in Biology*. New York: Simon and Schuster, 1979.

May, Robert. "Simple mathematical models with very complicated dynamics." *Nature* 261 (1976): 459–467.

2

How to Create Simple **STELLA**® Models to Solve Basic Equations

Everything I have ever done has converged to become system dynamics.
Jay W. Forrester

Biomarkers such as levels of cholesterol in the blood plasma, triglycerides, glucose, urinary creatinine, and anthropometric measures such as the ratio of weight to height, body composition, and body mass index represent keys to monitoring human health and assessing risk for disease. For this reason, educators and counselors in medicine, nutrition, and public health employ many formulae and quantitative relationships that provide means to assess health status and plan corrective actions or therapies for their patients. This is true not only for patients with established heart disease, diabetes, or obesity, but also for individuals who are interested in preventative medicine or personal strategies for maintaining fitness and minimizing the effects of aging.

Clinical values and quantitative relationships present a dilemma; they are of little value unless they are used, because most people are visual learners. People are interested in relationships. Without an appropriate tool to take the tedium out of performing calculations, the empirical relationships that have been identified in important clinical trials and biomedical research are overlooked, and their implications may be misunderstood. The personal computer can reduce the pain of dealing with quantitative data, and has the potential of making the mathematical tasks enjoyable. The computer feels no pain. A student who can learn to ask the computer to do his or her bidding has acquired the capability to perform tasks that are otherwise very difficult, if not impossible, for most people. And better still, this intermediary can lead to insights and understanding that are beyond one's grasp until interrelationships are clearly perceived. The student who doubts this potential for discovery may wish to reflect on how it must have felt for Albert Einstein to grasp the relationship between energy and matter, or for Keys and Parlin (1966) to derive an empirical relationship between dietary fats and serum cholesterol. When used in the best possible way, numbers can lead to a sense of awe, or aid in testing important hypotheses. From this viewpoint, it seems strange that the average person regards mathematics as an impediment to knowledge. How can this sense of frustration be overcome, so that students can work as freely with numbers as they do with verbal or visual images?

The aim of this text is to provide students with a means to make quantitative predictions about biological systems, and this cannot be done without learning how to use the best available tools. Some students are willing to carry out calculations, but many more avoid this task, perhaps remembering the tedium of long division. Would it not be worthwhile to set up simple computer programs to do repetitive calculations and eliminate math phobia? New computer programs display tool kits on the screen that may be used as an exceptionally simple and even fun method of creating customized computer programs. The screen displays a "graphical user interface" that permits the student to create programs, carry out automatic calculations, and display results as tables or graphs. A prototype of such programs is **STELLA**®, which can be used by anyone to create simple "machines" that calculate the results of any formula, or to explore the dynamics of complex systems over time. This text explains how to use **STELLA**® in the context of health assessment and biological dynamics, but any program with a graphical user interface and the ability to model systems may be used; other available programs are described in Chapter 26, "The Biokinetic Database."

2.1 Using **STELLA**® for Computer-Assisted Simulation

Professor Jay W. Forrester explained that one of the earliest programs for performing dynamic analysis was created by his colleague, Richard Bennett, in response to a request to encode equations for simulations to be run on their computer. The simulations were needed to complete an article on industrial dynamics in 1958 for the *Harvard Business Review*. Instead of generating computer code, Bennett made a compiler that generated the computer code automatically. This was the beginning of what came to be called the DYNAMO compilers, which ran on main frame computers and were the antecedents to **STELLA**® and more recent programs.

The current standard for computer simulation software includes a graphical user interface, which is represented on the computer screen when the program opens. It depicts a blank screen on which a system can be diagrammed using a menu bar and tools displayed in positions that enable operation using a mouse or other pointing device. The **STELLA**® program provides three levels of operation; the topmost *High Level Map* enables a global diagram to be created to help students use existing models without having to view the details of the system diagram or the equations that operate the program. However, the middle *Modeling Level* is the site on which a detailed map of the system is created and filled in with values that are used by the software to interactively generate a specific application program. The *Equations* may be viewed by entering the lowest level using the down arrows shown on the right side of the interface. These equations are called *finite difference equations*, and the master simulation software allows choices of methods for solving them. The methods are based on sets of equations called algorithms, which operate behind the scene. Many software packages for computer simulation also enable the user to enter experimental data pertaining to the system of interest, and to extract values from the data that provide the best fit to the model. **STELLA®** does not do this, but SAAM II®, MADONNA, and PowerSim® have this capability. Some newer spreadsheets such as the Excel® program provide curve-fitting routines, and the variety of math functions in the libraries included with such software permit the advanced user to perform simulations.

2.2 Tools Used to Create and Run Models with **STELLA®**

STELLA® is used to create blank templates for dynamic models in the same way that word processing programs create individual text documents. When the program is first opened, a menu bar and a set of tools are displayed along with a blank workspace. Unlike word processors, **STELLA®** files contain three levels of organization. When a new file is first opened, the top level is displayed, which is intended to provide a global diagram of the "big picture," and also allows files to be operated without reference to the details that actually run it. This may be compared to operating an automobile without having looked at the engine. However, most model development will make use of the next lower layer, where model diagrams are created and linked to the algebra that actually runs the program. To get to this layer, notice that the left side of the diagram always contains arrowheads pointing up or down. These are navigation tools used to move between levels in the model. Figure 2.1 depicts the tools on the menu bar in the middle layer. The menu bar along the top side of the interface allows for standard file operations (open, close, save, and print). It also allows the model to be run by choosing this option from the Run menu. Open each of these menu items and examine the range of choices allowed.

The majority of operations for creating models, however, make use of the tools displayed on the tool bar. These will be used immediately to explain general procedures. The key tools correspond to the icons labeled Stock, Flow, Converter, and Connector, for these are the building blocks that will be

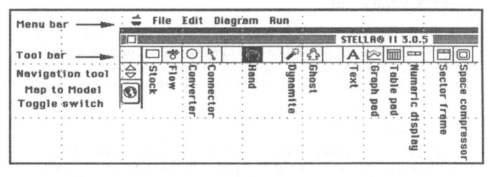

FIGURE 2.1

used to create models. They also correspond to four important properties of all systems. Each of these elements is chosen by using the *Hand*, a form of the cursor that is used to select and move icons on the system diagram. A *Stock* (shown by the rectangle) represents a reservoir or pool of material that is important for the model. Examples of stocks could include all the money in your bank account, all the milligrams of glucose in your blood, or all the grams of fat in your abdominal fat depots or hips. Stocks are identifiable objects that may be created or destroyed, but they are relatively stable quantities. Flows, on the other hand, represent rates of change. *Flows* (arrowhead and pipe) may be thought of as conduits or pipes or faucets and drains, but the key aspect of flows is the ability to move material into and out of stocks. If a stock represented money, then flows could represent monthly paychecks or cash outlays; if the stock represented abdominal fat, then flows could be the quantity of fat taken in from the diet over the course of a day, or the amount oxidized by muscular activity. Flows must have units that equal an amount of change in the material found in the stock over a specified period of time; if fat is measured in grams, then a flow could be grams per day, for instance. But how are flows to be controlled? Every pipe must have a valve or spigot to control the rate of flow, and this is the function of the *converter* (depicted as a circle). Converters may be used to indicate when the flow takes place, and how much material moves through the conduits into or out of the reservoir. The fourth key building block is the *connector*, an arrow that is used to indicate proportionality or a necessary input. For example, if a rate of flow equalled a rate constant times the amount of material in a stock, a converter specifying the rate constant could be linked to the flow with a connector. In figure 2.2 each converter is linked to another by a connector. The only other tool needed to get started is called the *dynamite* icon. Even dynamic modelers make mistakes, but these can be eliminated painlessly by selecting the dynamite tool and exploding the offending icon. Now let us open the program and create a model.

2.3 Lesson 1: Use **STELLA**® to Create a Calculator for Body Mass Index

After the mouse button is clicked, the display will show a file folder with a series of symbols (icons) that can be used to open different programs. Locate the **STELLA**® icon and click on it to start the program. The tools that will be needed will be described briefly and then used to create a working model.

The Menu Bar

A menu bar will be displayed at the top of the screen that permits the programs completed to be saved in the temporary save area of the hard drive or

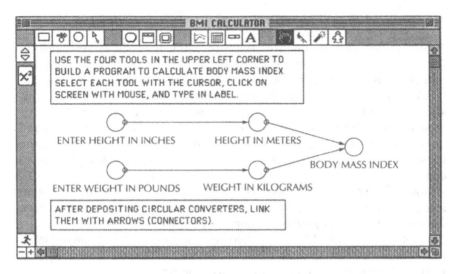

FIGURE 2.2

on a floppy diskette. If you wish to save your work, you should take a 3.5 inch diskette to the laboratory, and format it for the Macintosh (this will be done when you insert the diskette after a question is displayed asking whether or not you wish to format the diskette).

Figure 2.2 shows the screen with menu bar and work space as they will appear shortly after beginning to work with the program. Completed programs are run using the *Run menu*; more on this later. Menu bar operations allow you to work with your programs or files, and are now very similar for many different programs. To quit using the program at any time, select Quit from the File menu.

The Tool Bar

Below the menu bar is a series of symbols representing tools that are used to construct and run the programs. The important ones for this lesson, listed from left to right, are the symbols called the stock, the flow, the converter, and the connector. The cursor is shown as a hand symbol, and the firecracker is the dynamite tool that lets you remove any unwanted item from your program. Next to the letter A are a graph symbol and a table symbol. These two icons are used to view the results of your calculations.

Two other devices that will be used here are the up and down arrows located on the left side, which move to other levels of the program, and the earth/chi square button. This switches from mapping mode to math mode. The mapping mode allows the diagram to be drawn and the math mode allows you to enter numbers, equations, and relationships that complete the program.

Create and Use a Body Mass Index (BMI) Calculator

To start creating a BMI calculator, select the circle icon (the converter) from the tool bar by clicking once. Deposit the symbol on the screen with a second click, and type in a label as shown. To use the tool again, press the Option key on the keyboard (to the left of the space bar), and click again. This always lets you reuse the last tool without moving back to the tool bar.

Next, select the arrow icon (connector), move the cursor so it is centered on one of the symbols, and hold down the mouse button while moving the cursor into the next icon. This is called "clicking and dragging." Do this to connect all parts of the diagram as shown in the figure.

Now switch to the math mode as shown in Figure 2.3. Note that question marks appear on each symbol. This means that the diagram does not yet have the numbers needed to create a computer program.

Numbers are entered by using dialog boxes associated with each symbol that contains a question mark. Click on the symbol labeled "Enter height in inches." The view shown in Figure 2.4 will appear.

Type in your height in inches. You can use the keyboard or the pad shown in the dialog box. Now close the box by clicking on OK (lower right side) and go to the symbol labeled "Height in meters." One inch is 2.54 cm. Select Required Input, "Height in inches," and use the keypad to multiply by 2.54. Divide this by 100 because there are 100 cm in a meter. Or you can just multiply by 0.0254; the result is the same.

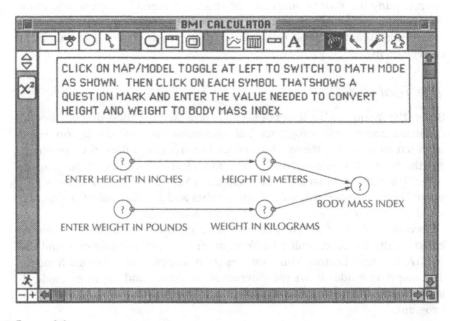

FIGURE 2.3

○ ENTER_HEIGHT_IN_INCHES

Required Inputs

E ( ) ^
7 8 9 *
4 5 6 /
1 2 3 −
0 . +
«

Builtins
ABS
AND
ARCTAN
CAP
CGROWTH
COOKTIME

○ ENTER_HEIGHT_IN_INCHES = ...

66

[Become Graph]   [Document]   [Message...]   [Cancel]   [OK]

FIGURE 2.4

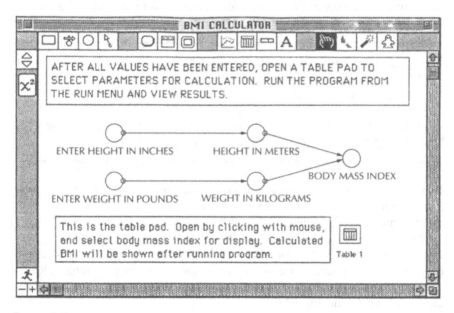

FIGURE 2.5

Close the dialog box and go to the symbol labeled "Enter weight in pounds." Type in your weight and close by clicking OK. Go to the symbol labeled "Weight in kilograms." Open this and select the required input, "Enter weight in pounds." Divide by 2.2 to convert pounds to kilograms. Click OK.

Time	ENTER HEIGHT IN INC	ENTER WEIGHT IN PO	BODY MASS INDEX		
.00	66.00	125.00	20.22		
.25	66.00	125.00	20.22		
.50	66.00	125.00	20.22		
.75	66.00	125.00	20.22		
Final	66.00	125.00	20.22		

BMI Table

10:05 1/11/96   Table 1

FIGURE 2.6

Body mass index is equal to weight in kilograms divided by height in meters, squared. Go to the BMI symbol, open the dialog box, and select "Weight in kilograms." Divide this by "Height in meters." To square this term, enter left and right parentheses, select the carat exponent (it looks like a hat) from the keypad in the dialog box, and then type the number 2. Click OK and close the box.

You have now completed a computer program. To verify your equations, click the down arrow in the left margin. It should show the following:

```
BODY_MASS_INDEX = WEIGHT_IN_KILOGRAMS/(HEIGHT_IN_
METERS)^2
ENTER_HEIGHT_IN_INCHES = 66
ENTER_WEIGHT_IN_POUNDS = 125
HEIGHT_IN_METERS = (ENTER_HEIGHT_IN_INCHES * 2.54)/100
WEIGHT_IN_KILOGRAMS = (ENTER_WEIGHT_IN_POUNDS)/2.2
```

To view results, select the table pad shown in Figure 2.6. Open this and select the parameters you want to view; then close the dialog box. Now go to the Run menu and run the program. What are your results?

The program you just created can now be saved and used again and again. Try checking your BMI if you lost or gained weight. To do this, just open the symbol for weight and enter a new value. The program automatically changes. Run the program again to get your new BMI. Does the outcome suggest you have any health risks?

Desirable Body Mass Index for Different Age Groups

The range of body mass index that is associated with the greatest longevity for men and women of different ages is listed in Table 2.1. Women with body mass indices below 17 have a higher incidence of amenorrhea than those

Table 2.1

Age (yr)	Body Mass Index (kg/m2)
19-24	19-24
25-34	20-25
35-44	21-26
45-54	22-27
55-64	23-28

within the desirable range. Individuals with BMIs above 30 have a higher incidence of symptoms associated with obesity, including elevated plasma glucose, poor sensitivity to insulin, hypertension, and gall bladder disease.

Finally, use the program to determine how much you would have to weigh for your body mass index to equal or exceed 29, which is the range in which health risks begin to develop. Many individuals in the United States have BMIs in the range of 30-40.

Quick Estimate of Weight Relative to Desirable Body Weight

Women should allow 100 pounds for the first 5 feet, and five pounds for each inch above 5 feet. For men, allow 106 pounds for first 5 feet, plus 6 pounds for every inch above 5 feet. It is permissible to add 10% for a large frame and to subtract 10% for a small frame. Set up a converter to express your weight either as a percentage of the desirable weight, or in terms of the approximate number of pounds that should be gained or lost to equal desirable weight. Does either of these methods suggest that you may wish to consider weight loss or gain to improve your health?

2.4 Lesson 2: Create a Device to Calculate Total Energy Content in Meals

Nutritionists assume that available food energy in calories can be calculated from the number of grams of protein (4 kcal/gram), carbohydrate (4 kcal/gram), fat (9 kcal/gram), and alcohol (7 kcal/gram). Use these values to create a device to automatically calculate total energy in a meal of known composition, and the fraction of energy from each constituent. This can be used to learn the contribution of fat grams to total energy intake, and as a tool to modify the diet by knowing how many grams of fat correspond to specified dietary goals. Use Figure 2.7 and corresponding equations as a starting point to create a model using the tools available in **STELLA®**: Open each circle (converter) that shows a question mark and type in the appropriate equation, given next.

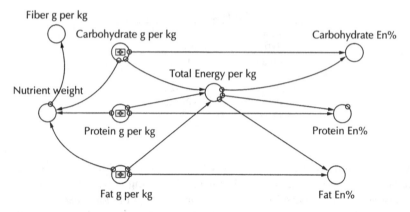

FIGURE 2.7

EQUATIONS FOR THE DIETARY ENERGY MODEL

Carbohydrate_en% = (Carbohydrate_g_per_kg * 4 *
100)/Total_Energy_per_kg
Carbohydrate_g_per_kg = 520
Fat_en% = (Fat_g_per_kg * 9 * 100)/Total_Energy_per_kg
Fat_g_per_kg = 100
Protein_en% = (Protein_g_per_kg * 4 * 100)/Total_
Energy_per_kg
Protein_g_per_kg = 160
Total_Energy_per_kg = (Carbohydrate_g_per_kg +
Protein_g_per_kg) * 4 + (Fat_g_per_kg * 9)
Total_weight = Carbohydrate_g_per_kg + Fat_g_per_kg +
Protein_g_per_kg

The equations just given above represent the name of each item named in the model (for instance, "Carbohydrate_g_per_kg" is the name of one converter) multiplied (asterisk, *) or divided (slash, /) by another value.

A starting point for dietary modification is to assume that relatively sedentary young women require a daily reference value of 2,000 kcal and sedentary young men require a value of about 2,600 kcal per day. Guidelines for dietary composition suggest a starting point of about 55% of calories from carbohydrate, 30% from fat, and 15% from protein. Enter these values and note the corresponding number of grams for each nutrient and for the total diet. In order to observe the output of your diet calculator, activate a table pad from the right side of the tool bar, place it on the work area, and open it by double clicking. Then select the names of the data that you wish to plot. Once you have done this, close the dialog box so that the graph or table remains open, and run the program. Now suppose that your personal dietary goals differ. Try adjusting the total number of calories to provide for a more active lifestyle at a lower fat intake. What dietary modifications would be needed to accomplish these aims?

STELLA® operates on three levels. The first level that opens can be used to run the program without viewing the system diagram. However, the Run Time version does not allow a user to add these devices, which are available in the Authoring version. The models that accompany this volume were created using this tool, and may be operated from the High Level Map. If the version you are using allows these devices to be added, create your own custom interface to run your program. For some models, such as the unit on bone remodeling, photographs and diagrams have been added to this level.

2.5 Modeling Projects Using **STELLA®**

Use **STELLA®** to perform the following calculations:

1. Set up an automatic device to convert English units to standard international units:

1 inch equals 2.54 centimeters
1 ounce equals 28.35 grams
1 kilocalorie equals 4.184 kilojoules. Convert kilocalories to megajoules (MJ).
Degrees Fahrenheit equal 9/5(°Celsius) + 32

2. Set up a calculator for resting energy needs from the equations of Harris and Benedict:

For men:

Resting energy needs (kcal) = 66.5 + 13.8 * (weight in kg) + 5 *
(height in cm) − 6.8 * (age)

For women:

Resting energy needs (kcal) = 66.5 + 9.6 * (weight in kg) + 1.8 *
(height in cm) − 4.7 * (age)

Now increase this by an activity factor to estimate total daily energy needs. Assume that the factor is about 1.2 for sedentary individuals, 1.4 for light activity, 1.6 for moderate activity, and 1.8 to 2.0 for heavy activity.

3. Modify the calculator for daily energy needs to estimate your daily requirements for fat in terms of percent of energy and grams of fat. Most advisory groups suggest that daily fat intake should not exceed 30% of energy needs, whereas about 15% should be protein, and the balance should be carbohydrate.

For a person who drinks alcohol, how would you modify the program to account for calories obtained from alcohol? What information might you seek in the medical literature that would assist you?

2.6 Record Keeping and Documentation

After you have completed a number of models, you may wish to review or modify your earlier work, or explain it to your instructor or colleagues. This can be rewarding if you have had an insight about how a system of interest really works, or frustrating if you cannot locate the appropriate version of the model and remember why you set it up in a specific way. Therefore, it is imperative to develop the habit of keeping records. The simplest means to remember the values used for any particular simulation is to print the list of equations, where the numbers and relationships used are shown. You may also save the equations as a text file using the File menu. Similarly, a copy of the system diagram may be printed when in the Model layer, using the Print Diagram command. The same holds for tables and graphs showing the results of any particular trial. It is highly recommended that these records be printed and stapled together, so that you can reconstruct your work. Another excellent idea for individuals who are interested in maintaining detailed records is to open a word processing program at the same time, and start a text file that allows one to keep a running record of progress. Current word processors allow text and images to be imported, so it is feasible to insert copies of your system diagram in this manner. To use this feature, first identify the kinds of graphics files that are compatible with your word processor, and save the images in this format. For example, PICT and GIF files are compatible with many word processors.

Use Standard International Units

Clinical values for biomarkers are usually expressed in Standard International units, and every person who works in health sciences must be familiar with this standard. For example, the beginning student should understand the relationship between temperature and heat energy (calories or joules), metric and English units of measure, and how to calculate body mass index on the basis of metric units of height and weight. In modeling, it is necessary and worthwhile to keep track of units so it is clear that a solution to an equation is expressed correctly. For more complex models, finding information expressed in proper units is often a challenge. Modeling can enhance familiarity with Standard International units, and also provides a simple way to interconvert familiar and unfamiliar units. Even if the clinician records blood pressure in millimeters of mercury, a student trying to solve Poiseuille's law of fluid flow will need to convert mm Hg to dynes per square centimeter by multiplying by 1,330.

References

Keys, A., and R. W. Parlin. "Serum cholesterol responses to changes in dietary lipids." Am. J. Clin. Nutr. 19 (1966): 175–181.

Internet Addresses:

One of the most exciting developments in computer modeling is the ability to share information and resources by electronic file transfer using the World Wide Web. This is an exploding area, but here are a few listings in system dynamics and modeling.

Centre for Nonlinear Dynamics in Physiology and Medicine, McGill University, http://www.cnd.mcgill.ca/. A group with research interests and training programs in biomedical system dynamics.

Directory of Biomedical Technology Resource Centers: ADVANCED COMPUTATION, http://www.ncrr.nih.gov/ncrrprog/btadvcom.htm. A directory of resources originally funded by the National Institutes of Health.

Library of Mathematical Models of Biological Systems, Georgetown University Medical Center, http://gopher.birc.georgetown.edu/model/home.html. A collection of mathematical models collected with support from the National Science Foundation by a prominent group of colleagues with special expertise in biomedical modeling and nutrition research.

MIT System Dynamics Group, http://web.mit.edu/sdg/www/. Lists home pages for several faculty and students, including Jay W. Forrester, John Sterman, and others. Tom Fiddaman's home page has a bibliography and list of models.

MIT System Dynamics Education Project, http://sysdyn.mit.edu/sd-intro/home.html.

National Simulation Resource at the University of Washington, http://nsr.bioeng.washington.edu/.

Pharmacokinetic and Pharmacodynamic home page, http://www.cpb.uokhsc.edu/pkin/pkin.html.

Resource Facility for Kinetic Analysis, http://www.ncrr.nih.gov/ncrrprog/btadvcom.htm#kin. Home of the SAAM II computer program, maintained by the group that developed and sells one of the major simulation programs used in biomedical research.

System Dynamics hyperlinks, http://www.sakasega.mgmt.waseda.ac.jp/fukushima/sd.html.

System Dynamics Mailing List, http://www.std.com/vensim/sdmail.html.

System Dynamics Modeling and STELLA on the World Wide Web (WWW), http://www.rtpnet.org/~gotwals/stella/stella.html. Extensive list of links and modeling projects maintained by Bob Gotwals.

Books

McArdle, W.D., F.I. Katch, and V. Katch. *Exercise Physiology: Energy, Nutrition, and Human Performance*. 3rd ed. Philadelphia: Lea and Febiger, 1991.

National Research Council. *Recommended Dietary Allowances*. 10th ed. Washington, D.C.: National Academy Press, 1989.

3

How an Equation from Physiology Can Become a Model

> I shall take the simple-minded view that a theory is just a model of the universe, or a restricted part of it, and a set of rules that relate quantities in the model to observations that we make.
>
> Stephen W. Hawking, *A Brief History of Time*, 1988, p. 9.

Students taking a first course in physiology or biochemistry encounter certain equations very early that typically evince a foreboding, if not outright terror. Memorable examples include the Nernst equation and the Henderson-Hasselbach equation; one remembers the *encounter*, not the content. And there were more to come: Poiseuille's law of fluid flow and Michaelis-Menten kinetics, the Gibbs-Donnan equilibrium; one could go on endlessly. Let me confess something right now: I knew the feeling of terror just as much as the next student, because I was never sure how to take a logarithm of a sodium concentration, or whether the intracellular concentration of ion was divided by the extracellular concentration or vice-versa, or why the electrical potential was negative and the action potential was depolarizing.

To begin to understand the capabilities one has in using the personal computer to solve equations, consider that in 1952, the electrophysiologists A.L. Hodgkin and A.F. Huxley used only a mechanical calculator in their Nobel Prize-winning work to obtain solutions to a complex differential equation that describes generation of an action potential.

The model included with this chapter was produced in a few minutes with **STELLA®**. It not only solves the Nernst equation, it converts it into a dynamic model. How much more comprehensible, and dramatic, to see the patterns emerging from the equation than to read and puzzle about the logarithm of the sodium concentration!

3.1 Let **STELLA®** Solve the Nernst Equation

Because cellular membranes are primarily made of lipid, they are semipermeable to ions and effectively separate charges. Further, by means of an energy-dependent process called active transport, they maintain gradients of ions such as sodium, potassium, and chloride that collectively generate a voltage across the membrane. This voltage is the membrane potential or potential difference, E_m. The voltage is the sum of the individual potential differences due to each ion whose concentration differs on either side of the plasma membrane. Transmembrane voltage may be calculated using the Nernst equation (Berne and Levy, 1993), in this case simplified for monovalent ions near body temperature:

$$\text{Potential difference, } E_{Na} = -60 \text{ mV} * \log_{10} \frac{(Na)_i}{(Na)_o} \qquad (3.1)$$

This is the voltage due to the sodium ion alone. $(Na)_o$ is the sodium concentration on the outside of the cell, and $(Na)_i$ is the concentration on the inside. To make the model as simple as possible, just assume that the membrane potential is the sum of the potential difference due to sodium and that due to the potassium ion:

$$\text{Membrane potential, } E_m = \frac{gNa}{(gNa + gK)} * E_{Na} + \frac{gK}{(gNa + gK)} * E_K \qquad (3.2)$$

The terms labeled g are conductances for each ion, and multiplying by these ratios gives a weighted average that predicts the equilibrium or resting membrane potential. The conductances are not fixed, but vary depending on the instantaneous voltage and presence of neurotransmitters at synaptic junctions. It is the change in conductances that actually generates the action potential, or membrane depolarization. Now here is a splendid aspect of the computer: **STELLA**® will take these logarithms, so all one needs to do is place icons on the mapping level in an appropriate way, enter some values, and the program will solve the equations. The user may then focus on the *meaning*, rather than the *mechanics*!

Here are some values that will allow us to create a working model: Sodium ion concentrations on the inside and outside of cells are about 10 mM and 120 mM, respectively; for potassium, the values are about 140 mM and 2.5 mM. Sodium concentration is high outside of cells, and potassium ion concentration is high inside of cells. For a simple model, all we need to know is the ratio of the conductances, because the factors are ratios. Typical values are that potassium conductance, gK, is ten times higher than sodium conductance when the membrane is at rest, but the sodium conductance increases tenfold when an action potential occurs. The potassium conductance increases somewhat to offset the sodium conductance and allow the cell to restore its resting potential quickly. Figure 3.1 shows the map of the model and a typical result, along with the equations.

To convert the equation into a dynamic model, two filips were added. Rather than just entering a fixed number for the conductances, the **STELLA**® Built-In function called Pulse was used to change the sodium and potassium ion conductances at intervals one may choose. Pulse is a test function that takes the form: Pulse (pulse volume, time of first pulse, duration between pulses). Action potentials often occur over milliseconds, but let us assume we are dealing with a cold-blooded turtle that has no need for speed; let our

E Na	72.61
E K	−104.89

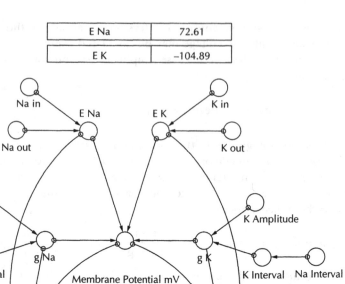

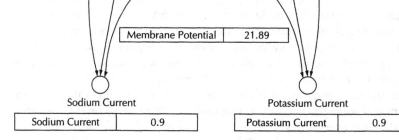

Membrane Potential	21.89

Sodium Current	0.9

Potassium Current	0.9

FIGURE 3.1

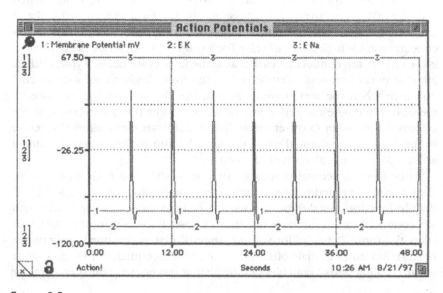

FIGURE 3.2

time units be seconds. Typical output is shown in Figure 3.2. Note that the action potentials begin from a resting membrane potential near the equilibrium value for potassium and move toward the resting value for sodium ion during the action potential.

The specific values for ion concentrations and resting membrane potentials vary among animals and different kinds of cells. A typical equilibrium potential for the potassium ion is near -90 mV, and for sodium, it is near $+60$ mV.

Open the model and run the program from the Run menu; feel free to change variables and observe outcomes. Use the sliders on the High Level Map to change the sodium concentration, the potassium concentration, the conductances, and the pulse intervals. Plot different variables to see how each contributes.

Is it easier to understand the Nernst equation with the help of **STELLA®** than when you first encountered it in your physiology text?

Accept this challenge: Use **STELLA®** to solve any equation from your physiology or biochemistry text. If the simple version of the Nernst equation is too easy, try the Constant Field equation or the Hodgkin-Huxley equation. An interesting example from biochemistry might be to set up a model that titrates the pH of solutions of amino acids, for instance. One could do this by using the calculation interval to add increments of hydrogen ions, rather than increasing simulation time by increments of DT.

EQUATIONS FOR THE MODEL OF THE NERNST EQUATION

```
E_K = -60 * LOG10(K_in/K_out)
E_Na = -60 * LOG10(Na_in/Na_out)
g_K = 10 + PULSE(K_Amplitude,6.5,K_Interval)
g_Na = 1 + PULSE(Na_Amplitude,6,Na_Interval)
K_Amplitude = 10
K_in = 140
K_Interval = Na_Interval
K_out = 2.5
Membrane_Potential_mV = E_K * (g_K/(g_K + g_Na)) + E_Na
* (g_Na/(g_Na + g_K))
Na_Amplitude = 10
Na_in = 9
Na_Interval = 6
Na_out = 120
Potassium_Current = g_K * (Membrane_Potential_mV -
E_K)/1386
Sodium_Current = g_Na * (Membrane_Potential_mV - E_Na)/
- 1386
```

References

Berne, R.M., and M.N. Levy. *Physiology*, 3rd ed. St. Louis: Mosby Year Book, 1993.

Hawking, Stephen W. A *Brief History of Time From the Big Bang to the Black Hole*. London: Bantam, 1988.

Hodgkin, A.L., and A.F. Huxley. "A quantitative description of membrane current and its application to conduction and excitation in nerve." *J. Physiol*. (London, 1952) 117: 500–544.

4

Rates of Change

[The girls of Copenhagen] also rode bicycles everywhere, and this, according to a physical law discovered at the [Niels Bohr] Institute, explained why so many students married Danish girls. "When girls are riding bicycles," the law stated, "you see more per second."

Barbara Lovett Cline

Prospects for matrimony notwithstanding, it is axiomatic that *time is of the essence*, even more in biology than in law and real estate. When journalists refer to the ticking of the biological clock, they usually mean that the choice between career and family is difficult for young women, because fertility declines after the age of thirty more rapidly in women than in men. However, one of the most amazing features of human biology is that important, even crucial, processes occur within vastly different time frames. Our senses must enable us to respond within milliseconds to changes that may threaten us, the secretion of insulin may change within minutes of taking a meal, and the expression of our genes may be reprogrammed over a few hours or less. Other cycles involved in response to sunlight and darkness take place over a day, while reproductive cycles tend to require several days to a month, and reproduction itself may require a significant part of a year. One of the most important aspects of cellular physiology, the cell cycle, requires about a day to complete in mammals, but only 20 minutes or so in microbes. If a *Salmonella typhimurium* ever challenges you to a race to complete cell division, decline the invitation, for mammals can't win!

For all these reasons, the most important concept that can be derived from the study of biodynamics is an idea of the way a system of interest changes over time. Before anyone can begin to think about time itself—and that is the Gordian knot of all problems in biology—one must have a way of representing change over time. The concept is so important that this chapter is devoted to explaining how material represented as a "single compartment" changes over time, and the connection between rates of change and a derived value known as the *half-life*.

4.1 The Steady State

The steady state is defined as a condition under which there is no net change in a system of interest. This is a point of equilibrium or balance. In reference to the single-compartment model, it indicates that the amount of material that enters the compartment is identical in quantity to the amount that leaves during any specified period. Yet the steady state does not indicate a condition of stasis or stagnation—it only means that for every unit of material that is eliminated, an identical amount is formed. In modeling, the steady state or equilibrium is important because it represents the desired starting point, from which a change in a variable of interest will produce effects that can be understood. The steady state is somewhat idealized, because it is unlikely that a true steady state ever exists—if the steady state existed, so might perpetual motion, perfect beauty, and immortality. However, one would like to set up a model in such a way that no change is occurring, and this can be achieved in the perfect world of mathematics, if not in the world of the experimental scientist.

For a single-compartment model with zero-order input and first-order elimination, the steady state exists when the input rate exactly equals the output rate, which is the product obtained by multiplying the elimination rate constant by the quantity of material present in the compartment at a specific time. This is our starting point for understanding change over time.

4.2 Temporal Behavior of the Single-Compartment Model

Now imagine that either the input rate or the output rate constant in a single-compartment model has been changed, either by increasing it or decreasing it, so that the system is no longer in a steady state. The result will be that the quantity of material in the compartment begins to change. Open the single-compartment model and compare the following conditions:

1. First set the input rate to 1.0, the output rate to 1.0, and the amount of material in the compartment (stock) equal to 1.0. Set the time specifications to 24 h, and run the model. A straight line should be obtained.
2. Change the input rate to 2.0 and run the model again. Note the time when the material attains a new steady state. Repeat this with the input rate set to 4.0. Compare the times when the new steady state is achieved.
3. Do the same experiment after changing the output rate to 0.5 and 2.0. Now do the times differ? Can you draw a general conclusion from the outcomes?

4.3 The Concept of Half-Life and the Time Course of Change

Two conclusions should be suggested by the previous experiment concerning a system characterized by a constant rate of inflow and an outflow rate that is a fixed proportion or fraction of the material in the compartment. First, changing the *inflow rate* to a different value causes the amount of material to seek a new level. However, the time required to attain the new level does not change; it is the same for any inflow rate. This is easiest to observe if one plots the amount of change as a percentage of the new steady state value. Then, 100% of the new steady state value will always be achieved at the same time. In a related manner, note that half of the total change also occurs within a fixed period. This result does not make sense to most people; it seems to suggest that it would take the same amount of time to increase the amount of material by twofold or by tenfold. Intuition suggests that there should be a difference, but the outcome belies this.

Second, changing the outflow rate by a similar multiple not only changes the level of material at the new steady state, it also changes the time required to attain a new steady state. The model suggests that the time required to attain a new steady state is not related to the rate of production of the material of interest, but to the rate of elimination. More emphatically, one could restate that the time required to attain a new equilibrium in a model characterized by fixed inflow and proportional outflow is not related to the rate of production, but to the rate of elimination!

4.4 The Concept of Fractional Elimination Rates

Suppose that the concentration of material in our compartment could be measured in terms of moles per liter; then the input rate would be moles per liter per hour (or other applicable unit of time). The output rate is the product of multiplying the concentration by a rate constant, so it must have units of per hour (h^{-1}). This should be boldly highlighted: *the output rate constant is expressed purely in terms of time units*. And what fundamental property of the system does it govern? The output rate constant specifies what fraction of the material exchanges during each time period. For example, if the output rate constant is small, it indicates that only a small amount of material leaves the system each hour. A value of 0.1 means that one tenth exchanges per hour (0.1 times the amount of material present leaves each hour). Similarly, a value of 0.5 h^{-1} means that half the material exchanges each hour, and a value of 2.0 h^{-1} would mean that the material is renewed every half hour. This value is commonly known as the *fractional rate constant*, and its implications cannot be overemphasized: **The fractional rate constant governs the rate of change** in this simple kinetic model.

This outcome would not be interesting unless the model applied to the real world, but it does apply in innumerable situations, and its implications must be understood, mulled over, grasped, and internalized until they become second nature to any scientist who is interested in time. Just consider for a moment that each cell in the human body contains several thousand individual and distinct proteins and messenger RNA molecules, each produced from a separate gene. Even though these molecules coexist in the same cells, some are extremely stable and persist for many weeks or months, whereas others appear and disappear within a few minutes even though chemically, all are comprised of the same constituent amino acids or nucleotides. One thinks of the incredible stability of DNA that has been mummified when an insect is caught in a drop of pine pitch, which turns to amber over many years.

Although there is nothing intrinsically unstable about these molecules, some exist but a moment and then are destroyed. The passage of time affects them differently; how can this be?

4.5 Half-Lives and Half-Times

The biological half-life of a substance is defined as the amount of time required for half of it to be eliminated, and this value is conveniently related to the fractional elimination constant. The relationship can be readily derived. Suppose the concentration of a substance initially equals C_0, and then declines to half of this value after a period of time defined as the half-life, $t_{1/2}$. The equation for change over time can be written and solved as follows, where C is the amount of material present at any time, t:

$$\frac{dC}{dt} = -k_e C_0 \tag{4.1}$$

Solving Eq. 4.1,

$$C = C_0 e^{-k_e t} \tag{4.2}$$

$$\ln C = \ln C_0 - k_e t \tag{4.3}$$

Rearranging,

$$\ln (C_0/C) = k_e t \tag{4.4}$$

After one half-life has passed, $C = 0.5\,C_0$, therefore:

$$\ln (2.0) = k_e t_{1/2} \tag{4.5}$$

$$0.693 = k_e t_{1/2} \tag{4.6}$$

$$t_{1/2} = 0.693/k_e \tag{4.7}$$

For this reason, when the fractional elimination rate equals $0.693\ \text{h}^{-1}$, the substance in the compartment will have a half-life of 1 h. Observe that the product of the half-life and the elimination rate constant always equals 0.693, the natural logarithm of 2.0.

Unless the reader has a natural ability to understand math, this outcome is much easier to demonstrate using **STELLA®** than it is to try to explain. But what does it mean? What should a person try to remember, since the mind is probably already preoccupied with one's spouse's birthday? To take our watchword from a song in the movie, *Casablanca*, "You must remember this": **The half-time for change in a single kinetic compartment is determined by the elimination rate constant.** Or, to make the statement even more general in case there is more than one route of disappearance, the half-time for change is governed by the sum of all the rate constants that apply.

The astute student who is not yet somnambulent may ask, "So, what happens if the input rate is not fixed, but is proportional to something else, such as when we have a protein being synthesized by translation of a messenger

RNA that in turn is produced by splicing of a nuclear precursor?" At which the able mentor would cheer, pat the student on the back, and suggest, "Why don't you set up a **STELLA**® model and find out?" You will begin to understand that you can use this system to perform true experiments and obtain valid answers, and that is an even more desirable outcome than knowing the answer to a single question.

There is no need to be impertinent. The answer to the astute student's question will turn out to be that the rate of change is really governed by the sum of all the time constants that apply.

4.6 Symmetry and a Helpful Rule

The kinetic behavior of material representing a single compartment is symmetrical in the sense that the time course of increase and the time course of decrease are the same after a specified change in input rate. For any unique value of the elimination rate constant, the change to 50% of the new steady state concentration will take place in one half-life. It is the same, no matter if the input has been increased twofold or decreased to one-half. Try this using **STELLA**® and see.

Now how long should it take to attain a new steady state? Here, an actual rule of thumb might be useful, so hold up your hand and picture a bathtub draining with a specified half-time, let's say one minute. After one minute, how much water is left? You should have answered, 50% or half, as you held up one finger to mark one half-life. And after two half-times? Half of 50% is 25%, so hold up one more finger. When you come to your thumb, five half-lives will have elapsed, and you should be down to 3.125%. So 97% of the change has occurred at five half-lives, and yes, I just counted this off on my hand the same way you did! People who wish to be rather exact will usually say it takes six half-lives. And whenever someone asks you in the future, you will know what to do even if you forgot the number. In fact, you will never forget this rule of thumb, so you just got your money's worth from this book! Try this question at the next graduate student mixer; I guarantee you will find at least one full professor who does not know how many half-lives must pass before a process can be considered to be complete!

4.7 **STELLA**® Can Make e Your Temporal Assistant

When my calculus book stated that the value of x for which ln x = 1 is denoted by the letter e, I did not pay much attention. That was a mistake on my part, because e turns out to be a very useful function, no doubt related to the fact that it occurs in the solution to differential equations like the ones previously shown. Yes, e equals 2.718, and so on, and ln e = 1. You will encounter this function innumerable times if you ever try to understand biological kinetics, so let us use **STELLA**® to find out why this function can be useful.

When you open one of the circular converter icons in the **STELLA**® program, you will observe a list of mathematical functions, among them one labeled EXP( ). This function is e raised to a power. If the exponent is negative, that means the function is a reciprocal, 1 divided by e raised to a power. If you scroll down the list of functions, you will also find one labeled TIME. This is simulation time, but it represents the same thing as t in the equations previously listed. So if you enter the rate constant, k*TIME, you can solve the riddle of the powers of e as a function of time and the rate constant. The model shown in Figure 4.1 was created to do this.

The following equations were entered in the dialog boxes of the converters shown in the figure to solve the exponential function of e raised to the power of k times time; the converters on the right side of the figure generated the reciprocal and 1 minus the reciprocal function.

```
Rate_of_Formation = TIME
Exponential = EXP(Fractional_Change_Rate *
Rate_of_Formation)
Fractional_Change_Rate = 0.693/Half_Life_hours
Half_Life_hours = 1
One_minus_Reciprocal = 1 - Reciprocal
Reciprocal = 1/Exponential
```

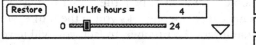

Restore	Half Life hours =	4
	0 ⊶▮⌁⌁⌁⌁⌁⌁⌁⌁ 24	▽

Fractional Chang...	0.17
Reciprocal	0.02
One minus Recip...	0.98

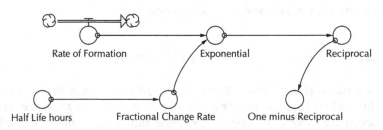

Rate of Formation   Exponential   Reciprocal

Half Life hours   Fractional Change Rate   One minus Reciprocal

FIGURE 4.1. **STELLA**® can solve equations that demonstrate how half-life is related to a fractional rate of change by way of the exponential function. When the half-life in hours is entered into the slider in the High Level Map, the Model converts it into a fractional rate of change (k). The exponential function than changes from a value of one to a value of zero as a function of time.

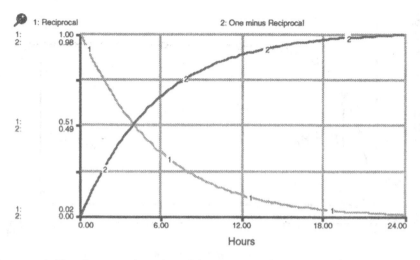

FIGURE 4.2. The change in the value of the exponential function, e^{-kt}, is shown during a simulated period of 24 hours for material with half-life of 4 hours. The exponential begins at a value of 1.0 and decreases to 0.5 after 4 h (curve 1). The function for fractional change, $1 - e^{-kt}$, begins at a value of zero and increases to 1.0. The student should vary the half-life in the model and observe the predicted fractional rates of change.

As time increases, the function, e^{-kt} and $1 - e^{-kt}$ produce the curves in Figure 4.2. The outcome shows that e^{-kt} is a curve that begins at 1.0 and declines to a value of zero when the product of time multiplied by the elimination constant is large. Subtracting this function from 1 produces $1 - e^{-kt}$, a curve that begins at zero and increases to 1.0.

What is the use of these two functions? First, observe that the time course of change depends on the magnitude of k. If k is large, then the functions reach their maxima or minima quickly. On the contrary, if k is small, much time must pass before the functions attain their full values. These exponential functions are the solutions to the single-compartment kinetic model, which have wide applications in pharmacokinetics, enzyme induction, and other time-dependent processes. Let us give an example.

Consider a messenger RNA that is present at a certain concentration, C_0, initially (at $t = 0$), but that increases to a new, steady state value as time progresses. If modeled as a single-comparment system, the induction curve takes the form shown in Figure 4.3. This graph represents one solution to the following equation for mRNA as a function of time:

$$R_t = R_0 + (R_{ss} - R_0)(1 - e^{-k_e t}) \qquad (4.8)$$

The concentration of mRNA at any time is designated R_t. The amount present intially is indicated by R_0, and the total amount of increase is given by subtacting R_0 from the amount present at the new steady state, R_{ss}. By multiplying the total change times the exponential function, $1 - e^{-k_e t}$, the solution for concentration at any moment is obtained.

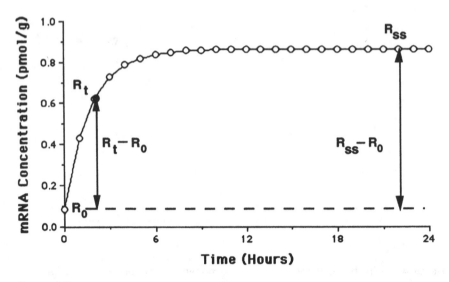

FIGURE 4.3

Now it is worthwhile to consider something that is equally interesting. Note that all that was needed to achieve the same behavior using **STELLA**® was to enter a rate constant for proportional decay in the dialog box. You did not need to enter Eq. 4.7 or any exponential functions; **STELLA**® solved the problem, and all you had to do was indicate the relevant rate parameters. This is too good to be true—you have acquired the ability to solve an ordinary differential equation painlessly, just by clicking and dragging icons across your computer screen! You are already solving difficult problems that indeed require calculus, and this is only the beginning of the feats you will be able to achieve with this program. Lest someone feel insulted, it should be mentioned that what you are doing is using a technique called numerical analysis to solve finite difference equations. If plus-or-minus 2% accuracy is good enough for you, please join me in doffing my hat to Dr. Jay W. Forrester and his colleagues.

4.8 Use Powers of e to Simulate Time-Dependent Events

Suppose it becomes necessary in a model to turn a switch on or off, or change a function over a specific period of time, so that it begins with one value and gradually changes to a new value. Powers of e allow any variable that changes in this manner to be simulated without adding another stock-and-flow to a system. Multiplying any number times $1 - e^{-k_e t}$ yields a function that begins at zero and changes until it equals the value of the multiplier, and the opposite effect is obtained by multiplying times $e^{-k_e t}$. To implement this in **STELLA**®, simply open the Exp function in the Built-Ins menu and substitute the Time function from the dialog box for t in the equation. The function

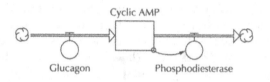

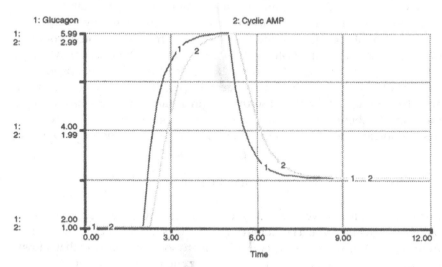

FIGURE 4.4

would look like this: $1 - \text{EXP}(-0.693 * \text{TIME})$, where 0.693 is the rate constant that gives a half-time of 1 time unit. An example is shown in Figure 4.4 using a Step function with this feature embedded. One point to watch, however, is that if one begins when the value of Time is not zero, the Time function will take on the current value of the simulation time. To make the function begin at zero, it would then be necessary to subtract the current value of Time (for instance, a Step programmed to occur at 2 h might contain an exponent written as follows: $\text{EXP}(-4 * \text{TIME} - 2)$. This procedure is a bit cumbersome, but it allows one to turn input rates on and off as desired.

```
EQUATIONS FOR STEP FUNCTION MODEL OF CYCLIC AMP

Cyclic_AMP(t) = Cyclic_AMP(t - dt) + (Glucagon -
Phosphodiesterase) * dt
INIT Cyclic_AMP = 1
Glucagon = 2 + STEP(4 * (1 - EXP(-2 * (TIME - 2))), 2)
- STEP(3 * (1 - EXP(-2 * (TIME - 5))), 5)

Phosphodiesterase = 2.0 * Cyclic_AMP
```

In a case such as this in which the input function is time dependent, the response has a time signature governed by both the input and output rate constants. Note that the Step function for glucagon has a time course dictated by the rate constant in the exponent (here, the value is 2). This means that instead of rising instantaneously to a new value, the step is gradual. Because the cyclic AMP has its own time course for elimination by the action of the phosphodiesterase, the accumulation curve is characterized by a delay between action of the stimulus and the response. This can be noted from the plot for the stimulus called Glucagon in the graph, which peaks and declines 1 minute before the cyclic AMP does. If this explanation seems opaque, try using **STELLA**® to understand the concepts! Open the converter for glucagon and change the Step functions so they begin at different times; also change the number subtracted from Time in each exponential function. The two values within each Step function must be the same for this procedure to work.

4.9 The General Function for Fractional Change

Just when you may have thought it could not get any better, let us consider one final generalization. The number $1 - e^{-k_e t}$ is just a fraction that ranges from zero to 1.0. We may now consider how to solve an equation that shows the fraction of total change that occurs over time in the single-compartment model. It is a completely general result and can apply to mice or molecules. The fractional change at any moment is given by subtracting the amount

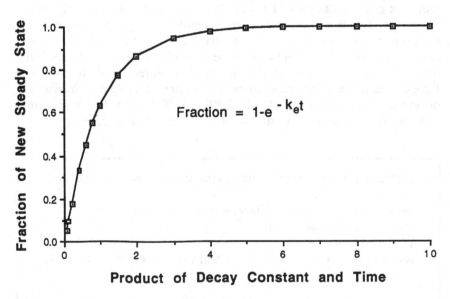

$$\text{Fraction} = 1 - e^{-k_e t}$$

FIGURE 4.5

of material present initially from the amount present at any later time, and dividing that amount by the total amount of change. When written as an equation, here is the general form, letting C stand for concentration:

$$f = \frac{C_t - C_0}{C_{ss} - C_0} = 1 - e^{-k_e t} \qquad (4.9)$$

How about that? Our old friend, the exponential function, is a simple way of indicating time-dependent, fractional change for any system that is characterized by first-order degradation or elimination—and, gratifyingly enough, that is true for the majority of biological molecules. However, let's look at a graph of this function (Fig. 4.5), just to be sure we understand. Let us plot the fractional change as a function of the product, k times t.

To conclude this brow-beating, it is useful to remember at least this one concept, because it introduces a very general method for thinking about change over time. Once you understand this idea, you can begin to think about time itself, and time-dependent processes, such as aging and a hundred other problems. And that is a nut that no one has cracked, so happy hunting! Though one might add, if you are starting to get the idea that time is not the same thing to a man, a woman, a child, a dog, and a goldfish just because they are all in a room with one clock, you may just be on to something—no matter how many times a second a cesium atom oscillates!

5

The Steady State:
A Question of Balance

If I ask you whether your brain is an equilibrium system, all I have to do
is ask you not to think about elephants for a few minutes, and you know
it isn't an equilibrium system.

Arnold Mandell (cited in Gleick, 1988, p. 298)

There is no perfect steady state, or condition of complete equilibrium. But
just as the invention of the number, zero, was necessary and profound, the
condition of the steady state is crucial to modeling. The steady state is considered to be a condition in which the total amount of material entering a system exactly equals the amount leaving the system; moreover, the amount of
material entering each defined kinetic compartment must exactly equal the
amount leaving that compartment. It is a condition of balance in which there
is no net change in mass over time.

The importance of the steady state is that it allows a model to be set up
with initial conditions that do not vary, so that the changes that do occur after the system is perturbed can be interpreted and understood. The purpose
of this chapter is to demonstrate how to set up a model in the steady state.
Incidentally, an equivalent condition is also important in most experimental
studies. For instance, if an investigator is monitoring a change in a cellular
property such as the concentration of a protein or mRNA, he or she must first
identify conditions under which the property is invariant, and this is not a
simple matter. Usually, one must ensure that a constant number of cells is present and that the concentration of substances in the medium does not change
(or at least, that it changes the same way for all samples) except for the variable of interest. Identifying such conditions is an art form, not only in modeling but in laboratory research.

5.1 Setting Initial Conditions for **STELLA**®
Compartmental Models

When models are first constructed, the user must determine how much material will be in each compartment and ensure that values are not changing.
One alternative is to enter zero as the initial value and let the computer calculate how much should be present. This can be noted using the Table Pad.
However, this procedure is time-consuming and there is a better way, which
will be explained so that the student can keep a record of values that were

used for successful trials. Then, when models are saved for future use, one can refer to the earlier work and not need to start from the beginning.

5.2 Calculating Initial and Final Values for STELLA® Models

The best method of determining amounts of material in kinetic compartments is to write a steady state equation that will permit the values to be calculated with a hand-held calculator. The methods that permit one to do this may be explained as follows:

1. When a system *is* at equilibrium, the sum of all inflows minus the sum of all outflows equals zero, and all the values are constant and unchanging. Another way of stating this conclusion is that the sum of all rates of synthesis equals the sum of all rates of degradation or transfer.
2. When a system *is not* at equilibrium, the extent of change in a value equals the sum of all inflows minus the sum of all outflows during each interval. This is the rule the computer program uses to calculate how the system will change over time on the basis of the finite difference equations.

Let us consider two basic models to explain how to obtain these crucial numbers.

Case 1: Single-Compartment Model with Two Outflows

This case is illustrated in Figure 5.1. Define the steady state concentration as C_{ss}. At steady state, the rate of synthesis equals the total rate of degradation and transfer:

$$\lambda_1 = (\lambda_2 C_{ss} + \lambda_3 C_{ss})$$

This also means that at steady state, the sum of all inflows and outflows is zero:

$$\lambda_1 - (\lambda_2 C_{ss} + \lambda_3 C_{ss}) = 0$$

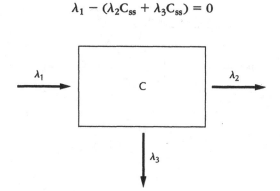

FIGURE 5.1

Rearrange the terms to be able to calculate the concentration, C_{ss}:

$$C_{ss} (\lambda_2 + \lambda_3) = \lambda_1$$
$$C_{ss} = \lambda_1/(\lambda_2 + \lambda_3)$$

At steady state, the concentration of material in a physical space equals the ratio between the sum of all inflows and the sum of all outflows.

Inflows have units of *material* synthesized or transferred *per time unit*, and proportional rate constants have units of 1/time, or time^{-1}. Mentally, go around the box in Figure 5.1 and add up all the material coming into the box. This sum will have units of material per time. If the example is mRNA, units for intake could be molecules per cell per minute (molecules cell^{-1}min^{-1}). At steady state there is no change, and we need to eliminate the time units. This is done by adding up all the rate constants, which have units of time^{-1}. We then divide by that sum. For example, if the input rate is molecules cell^{-1}min^{-1} and the output rate is a fraction per minute, then:

$$(\text{molecules cell}^{-1}\text{min}^{-1})/(\text{min}^{-1}) = \text{molecules cell}^{-1}$$

This is a concentration, which is expressed in the units we want. Now calculate the initial concentration, C_{ss}. Suppose that $\lambda_1 = 1$, $\lambda_2 = 0.5$, and $\lambda_3 = 1.5$, then

$$C_{ss} = 1/(0.5 + 1.5) = 0.5.$$

The scale may be chosen to contain this value, and the results of any changes may similarly be calculated. For instance, if the input is increased from 1 to 10, then the steady state value will be tenfold higher.

Let us construct a **STELLA®** model to prove this point. Assume that a nuclear mRNA precursor is being synthesized at a rate of 10 molecules per cell per hour, and that it is subject to nuclear degradation with a half-time of 0.5 h (rate constant = 0.693/0.5 = 1.4 h^{-1}). Assume that it can also be transported to the cytoplasm with a half-time of 1 h (rate constant = 0.693/1 = 0.7). Calculate the concentration, and run the model to verify your result. Try increasing the input rate, L1 (refer to the diagram in Figure 5.2) tenfold and

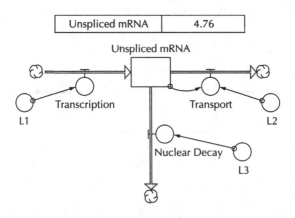

FIGURE 5.2

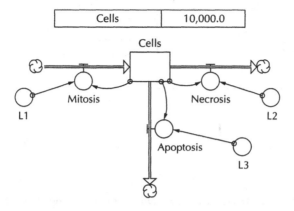

Cells	10,000.0

FIGURE 5.3

see what happens. Play with this a bit; try any combination and verify that you and your computer are in harmony!

What do you think would happen if one of the outflows was changed? The effect would be smaller; to get a tenfold change, *both* outflows would need to be changed by tenfold. Try this, and then contemplate what the result may mean about the effectiveness of nuclear mRNA as a control point for protein synthesis.

Now let us construct a minor variation on the theme by having an input rate that is proportional to preexisting material. For instance, assume that cells are formed by mitosis with a division time of 1 day, which gives a half-time for cell division of $\ln 2/1 = 0.693$ d^{-1}. Enter this value into the circle (converter) labeled L1 (see Figure 5.3). Assume that cells die by the processes of apoptosis (programmed cell death) and necrosis (a nonspecific response to a toxin or physical change), and that the rates are both proportional to the number of cells present. Arbitrarily assign these rates equal to 0.293 (L2) and 0.40 (L3). Now run the model and observe whether or not cell numbers change. Try other permutations, but by all means, try to avoid inducing a tumor! If you create a situation characterized by exponential growth, even **STELLA®** will not be able to keep track of all your cells! Compare your results with the previous outcome in which the input rate was fixed (or zero order).

Case 2: Two-Compartment Model with Exchange Between Compartments

This case is shown in Figure 5.4. This is more realistic than a one-compartment model, because we will normally include at least the blood plasma (or, formally, an "accessible" compartment we may sample) and other bodily organs (or their kinetic equivalents) as other compartments. According to this diagram, the following values will be found at steady state:

$$C_1 = (\lambda_1 + \lambda_4 C_2)/(\lambda_2 + \lambda_3) \tag{5.1}$$

$$C_2 = \lambda_2 C_1/(\lambda_4 + \lambda_5) \tag{5.2}$$

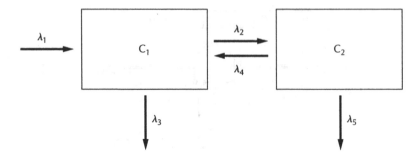

FIGURE 5.4

It is difficult to solve these equations without simplifying them. Note that all the lambdas are just constants. The complex set of constants in Eqs. 5.1 and 5.2 can be simplified by lumping constants together as follows:

$$C_1 = X + YC_2 \qquad \text{New (5.1)}$$

in which X equals $\lambda_1/\lambda_2 + \lambda_3$ and Y equals $\lambda_4/\lambda_2 + \lambda_3$.

$$C_2 = ZC_1 \qquad \text{New (5.2)}$$

in which Z equals $\lambda_2/\lambda_4 + \lambda_5$. Equation 5.2 is easy enough, it is just a constant times C_1. Equation 5.1 needs to be simplified and expressed in terms of just one value, C_1, instead of C_1 and C_2. Here is how to do this, as follows.
   Because C_2 equals ZC_1, insert this value for C_2 into Eq. 5.1:

$$C_1 = X + YZC_1 \qquad (5.3)$$

Now rearrange this equation so that C_1 is defined in terms of the three known constants:

$$C_1 - YZC_1 = X$$
$$C_1(1 - YZ) = X \qquad (5.4)$$
$$C_1 = X/(1 - YZ)$$

   The equations previously derived let us calculate starting values for the model shown, and the same method can be used for any model. Now finish by solving the equations using the parameter values in Table 5.1. The results

TABLE 5.1 Worksheet for a two-compartment model

Definitions:				
$X = \lambda_1/(\lambda_2 + \lambda_3)$		$C_1 = X/(1 - YZ)$		
$Y = \lambda_4/(\lambda_2 + \lambda_3)$		$C_2 = ZC_1$		
$Z = \lambda_2/(\lambda_4 + \lambda_5)$				
Parameter:	λ_1	λ_2	λ_3	λ_4	λ_5
Value	1	0.5	0.2	0.1	0.2
Factor	X =		Y =		Z =
	YZ =		1 − YZ =		
Amount	C_1 =		C_2 =		

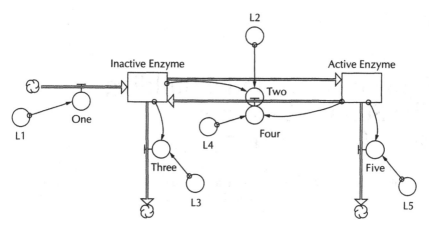

FIGURE 5.5

can be used to calculate numbers for the **STELLA®** program. When you have calculated these numbers, you are ready to run the program. Open **STELLA®**, generate a two-compartment model of the same design, and verify that initial conditions will be stable if these values are used. A graph will give a straight line.

To show that the same relationships will hold even after a change, try changing the input rate to 4 and verify that the equations give the correct value for the new steady state. If you calculate the numbers correctly, you will get exactly the same values with your calculator that you get using **STELLA®**! The advantage of **STELLA®** is that it predicts the time course of change for your system, which is something very few people can do without a computer.

A two-compartment model is shown in Figure 5.5 to help perform this exercise.

Once you have mastered this basic example, try working out initial conditions for a model of your own design that has a different structure. For instance, a three-compartment model is often the starting point for kinetic modeling. Use a system that is appropriate to the research project you plan to do.

6

Equations for Model Building: The Surface Law and Body Composition

A comparison...would suggest a trend of modern American animals to take the surface law less seriously than did the earlier European animals.
Max Kleiber, 1975, *The Fire of Life*, p. 202

6.1 Purposes for Modeling the Human Body

If our health depends on the balance between energy intake and expenditure, and moderate consumption of foods and appropriate medicines, then there are at least three kinds of model that could be useful to our attempt to understand how to maintain health and prevent disease.

1. Models of nutrient and drug kinetics, which focus on the absorption of substances, their movement into the bloodstream and distribution to tissues, metabolic effects, and subsequent oxidative destruction and elimination from the body. The highest level of development of such models is in the disciplines of pharmacokinetics and pharmacodynamics, but exactly the same ideas apply to requirements for nutrients and effects of hormones (Carson et al., 1983). Drugs are considered to be nonnutritive, exogenous substances that often act on receptors with specificity and are useful for treating disease, whereas nutrients are regarded as normal constituents of the body that are used to support growth and all metabolic processes.
2. Bioenergetic models examine the maintenance of bodily constituents and energy reserves, with the goal of understanding how sources of food energy maintain the major tissue types, and how the tissues change during training or weight loss. This kind of model is especially useful to exercise scientists and fitness specialists, who are familiar with the concept of Reference Man and Reference Woman (ICRP, 1975). However, this perspective also has utility for health care workers who attempt to prevent undue loss of lean tissue in patients who have been subjected to trauma, or who seek to prevent obesity or counsel patients concerning safe weight loss.
3. Growth models examine requirements for cell division and growth of the body or its individual components over time, beginning at the moment of conception or some suitable time thereafter. Periods of rapid growth that are important to understand include fetal development, the neonatal and adolescent growth phases, and the origin and treatment of cancers.

The close connection between the dynamics of drugs and nutrients is evident when one considers that safety and efficacy must be established for vitamins and minerals as well as for drugs and food additives. The confusion between the fields of nutrition and pharmaceutics is evident from the recent profusion of companies that sell nutritional supplements for body builders and alternative medicines, which often contain "natural" remedies for self-diagnosed conditions ranging from premenstrual syndrome to impotence. Even powerful hormones such as melatonin and dehydroepiandrosterone (DHEA) are now sold as "diet supplements" under our bizarre legal code. To the scientist, it is evident that all herbal preparations contain active substances that would be classified as drugs if they were purified. The distinction is unimportant in terms of principles of uptake and use, but extremely important to the governmental agencies that oversee the safety of foods and drugs.

6.2 Sources for Equations Used in Model Building

The process of model building requires both simplification and quantification. Simplification is the process of eliminating all items that do not directly bear on the problem of interest. For instance, the health care worker who is interested in treating obesity does not need a model of the body that is subdivided into all the individual organs, bones, and muscles. For many purposes, considering only fat mass and fat-free mass in a two-compartment model is sufficient. One can determine current fat mass by a variety of methods, such as by measuring body density by underwater weighing or by measuring thickness of skin folds with calipers, but how does one predict changes in fat mass over time? To do this, one must also understand that a particular change in food intake will produce a corresponding change in body fat, and the value most often used is 3500 kcal per pound. Later, we will examine this notion, but first, we consider the basic elements.

Where can the student find these quantitative relationships, in order to build or modify a model of personal interest? For this purpose, it is necessary to conduct research in the medical library, using computer-assisted searching for relevant terms. This should provide references to the primary medical literature, symposia, and published books in which results of ongoing work are summarized. The scientist is often able to obtain some of the necessary information by conducting research, but it is never possible to measure every variable that affects a system in a single study, and few students (or scientists) have the funds, lab space, or equipment needed to conduct sophisticated research.

Scientists always express their quantitative results in the form of numbers, and also use equations to express relationships if enough data have been gathered. Correlations between an independent variable (food intake, for instance) and a dependent variable (body fat) may be determined by collecting a large amount of data and applying the statistical procedure called regression analysis to determine an equation or number that shows typical quantitative relationships. These *regression equations* are very useful in model building, if one hopes to project typical changes seen in a group of people as a result of a change in behavior or a medical treatment. Statistical methods also provide a measure of population variability, so that the investigator can determine whether a particular subject is a good, average, or poor responder to treatment. A second kind of relationship can be expressed as an *allometric equation*. Allo- means different, and metric means to measure; allometrics is the process of measuring differential change. Allometry is most commonly used to monitor changes in organ size during growth. For instance, the head grows extremely rapidly relative to the body during fetal development and slower after the first year of life, so that babies have large heads compared to adults. This is allometrics! Another important example is the attempt to extrapolate drug doses that have been determined in one animal species to another species, including humans. For instance, if one determines how much of a carcinogen causes cancer in a group of rats, can one predict how much would be similarly effective in humans? An excellent summary of equations that have been derived in many studies of animal form, function, and lifespan can be found in Calder (1984).

6.3 Models of Human Body Composition: Reference Man and Reference Woman

The attempt to standardize or establish reference values is important in quantitative thinking. Just as we measure heights and depths and atmosperic pressure in comparison to conditions at sea level, the physiologist thinks of body composition relative to a standard Reference Man and Reference Woman. Reference Man is between the ages of 20 to 30 years, weighs 70 kg, and is 170 cm in height. His body is separated into five components: (1) fat and a lean tissue compartment that contains (2) total body protein, (3) total body water, and (4) minerals (or ash), and (5) approximately 0.15–0.5 kg of carbohydrate. The latter is often ignored because it is used and replenished daily, though this can give rise to weight fluctutations of up to 3 kg (6 pounds) because each gram of glycogen contains 2–3 grams of water. The average amount of glycogen stored is ordinarily about 0.15 kg, so for most people, this accounts for about one pound of weight fluctuation.

The concept of Reference Man was developed for Caucasian males in North America and Europe, and it was never intended to apply to populations throughout the world. Although readers who have different backgrounds may be interested in asking whether or not this idea applies to them, data from the Centers for Disease Control do not support the idea that growth potential is much different among different races—for instance, the tallest man in the world is presently a 7 ft, 9 inch (or 236 cm) North Korean basketball player!

More recently, Ellis (1990) reported data for 175 men and 1,134 women ranging in age from 20 to 90. The average white male in the 20-24 year group actually weighed 79.9 kg and was 1.78 m tall. The average white female in that age group weighed 59.2 kg and was 1.62 m tall. A small group of black men and women were the same height but on average weighed more (averages were 72.1 kg for adult women and 85.5 kg for adult men). For the white men, body mass index averaged 23-24 kg/m^2 and increased to a value of 26 by the age of 50; among 20-24 year old women, average BMI was about 22 and increased to 25-26 by age 50.

There are many idealized concepts of body shape and composition. Many individuals wish to gain muscle mass and develop larger, better developed bodies, whereas others focus on dieting and weight maintenance or fat loss as a desirable goal. Whereas these different views of weight and body composition are only indirectly concerned with the desire for good health, it is true that most people gain weight as they age to the extent that about a third of the U.S. population is now classified as obese (Thomas, 1995). Perhaps a lesson in modeling can help the student appreciate the fundamental difference between what is needed to gain or lose fat versus muscle! This has a great deal to do with health due to the numerous health risks of obesity, from diabetes to hypertension and gall bladder disease. And to the physicians and health care staff who work everywhere except the weight loss clinic, unusual weight loss is regarded as extremely undesirable and a very poor prognosis. Loss of lean tissue may be a sign of alcoholism, intestinal or kidney disease, hidden infection, cancer, or may result from burns, surgery, or trauma.

6.4 The Surface Law: Calculating Surface Area in Adult Humans

It takes no training in biology to answer whether a grown human produces more heat per day than a baby or a small animal; of course a larger body produces more heat. However, the question requires more thought if restated, "Does the adult human produce more heat *per unit body mass* than an infant or a small animal?" Any student who replied without hesitation that the adult human produces *less* heat per day per kilogram than an infant probably understands a well-known relationship. Although total body weight is positively correlated with metabolic needs, the amount of energy required per unit weight or mass decreases as mass increases. For over a hundred years, physiologists have attempted to explain this partly on the basis of surface area. The results have very practical consequences in explaining why infants become ill quickly when their water intake is not adequate, and why it is a very bad idea to lock your child or pet dog in the car during a summer shopping spree.

The surface law states that the basal metabolic rate is proportional to body surface area. Another way of stating a proportional relationship is that basal metabolic rate equals some constant (k) multiplied by surface area:

$$\text{Basal metabolic rate} = k \times \text{Surface area} \qquad (6.1)$$

If you place your fingers on your neck and then on any nearby wall, you will feel the warmth of your skin and the relative coolness of the wall. The warmth is due to heat being generated during metabolic reactions in lean tissues, which is carried to the blood and then to the body surface. The amount of energy we require in daily meals must offset the amount lost through the skin, and must be balanced so that our core body temperature may be maintained at about 37°C. Therefore, in order to think constructively about daily energy balance, it is useful to create a tool to estimate surface area. This will be used as part of a more complex model in the following sections on energy needs.

STELLA® Application: Create a Model to Calculate the Surface Area of the Human Body Using the Formula of DuBois (1916)

Open **STELLA®** and create the diagram shown in Figure 6.1. Into the dialog boxes associated with the converters, enter the equations shown next. This will allow estimation of surface area on the basis of body mass in kg and stature in cm. For comparison, the model also estimates the surface area of a cylinder of the same height for any radius desired.

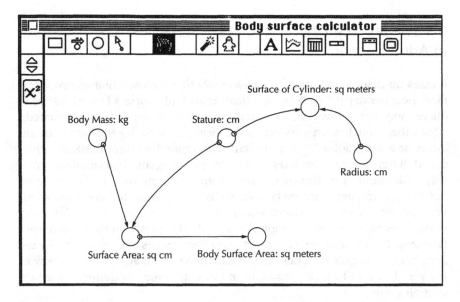

FIGURE 6.1

```
Body_Mass:_kg = 40
Body_Surface_Area:_sq_meters = Surface_Area:_sq_cm/10^4
Radius:_cm = 15
Stature:_cm = 165
Surface_Area:_sq_cm = 71.84 * (Body_Mass:_kg^0.425) *
(Stature:_cm^0.725)
Surface_of_Cylinder:_sq_meters = PI * 2 * (Stature:_cm
* Radius:_cm + Radius:_cm^2)/10^4
```

Can you use the Kleiber equation for metabolic rate shown in Table 6.1 to estimate the rate of heat loss from a person of the body mass you entered in your model? Alternatively, try making a model that incorporates the equation for heat loss as a function of ambient temperature. At any specified temperature, how much would the metabolic rate need to increase to maintain body temperature?

6.5 Understanding Exponents for Body Surface Area and Mass

Surface area (A) is obtained by multiplying two lengths such as the width and height of a rectangular object, and therefore has units of length squared (e.g., square meters, square centimeters, or square miles). Volumes (V) are obtained by multiplying by a third length, so that volumes are proportional to length cubed (l^3). Because the human body occupies a volume, it seems evident that weight or mass is also proportional to length cubed. For instance, a kilogram is defined as the mass of water contained in a cube that is 10 cm on each side, so that it contains 1,000 cm³.

$$A \propto l^2; \quad l \propto A^{1/2} \tag{6.2}$$

$$V \propto l^3; \quad l \propto V^{1/3} \tag{6.3}$$

If it is true that mass and volume are proportional, then it can be shown that area is proportional to body mass, based on expressions 6.2 and 6.3:

$$A \propto l^2 \propto (V^{1/3})^2 \propto V^{2/3}$$

Therefore,

$$A = kM^{2/3}, \tag{6.4}$$

where k depends on the shape and density of the body. For humans, an approximate relationship is that the surface of the human body averages about 12.3 square decimeters per unit of the 2/3 power of body mass in kg (Calder, p. 184). For slim people the constant would be higher, and for obese people it would be lower. DuBois and DuBois (1916) derived a related formula that allows more accurate calculations:

$$A = 71.84 * (M^{0.425}) * (L^{0.725}) \qquad (6.5)$$

where A is surface area in cm^2, M is body mass in kg, and L is stature in cm.

6.6 The Allometric Equation

For any physiological function that is correlated with body mass, the scaling or proportionality can be described relatively well by the following equation:

$$Y = aM^b$$

where Y is the variable in question, M is body mass in kg (or m is mass in g), a is a proportionality constant, and b is a scaling factor. If b = 0, $M^0 = 1$ and mass has no effect on the function of interest. If b = 1.0, there is a linear relationship. The following table lists a number of equations that pertain to the scaling of different components of the human body.

6.7 Use **STELLA**® Models to Solve Equations with Exponential Terms

STELLA® can easily help you solve equations such as Eqs. 6.4 and 6.5, so that you will be able to see a number or graph that explains the meaning of the equation. This can be useful, for example if you needed to calculate surface area for a client in your clinic, or if you were interested in human thermoregulation or environmental physiology. Let us make models that will solve the allometric equation and the surface area equation of DuBois (1916).

The equation $(Y = aM^b)$ is solved in the model shown in Figure 6.2, where a is any constant, and b can be any negative or positive integer or fraction, entered as numerator divided by denominator:

```
Calculate_Y =
Enter_the_proportionality_factor_a *
Enter_a_value_for_M^Enter_any_exponent_b
Denominator = 3
Enter_any_exponent_b = Numerator/Denominator
Enter_a_value_for_M = 4
Enter_the_proportionality_factor_a = 1
Numerator = 1
```

Solve the equation,
Y = aM^b, where b = Numerator/Denominator.
The carat (^) designates that M is raised to
the power of b (which may be negative or
positive in sign).

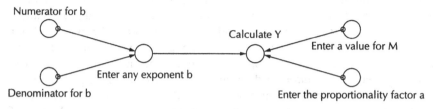

Numerator for b

Calculate Y

Enter a value for M

Enter any exponent b

Denominator for b

Enter the proportionality factor a

FIGURE 6.2

Practical Application: Energy Needs and Body Mass

Kleiber (1932 and 1975) estimated that the basal energy needs for a female rat weighing 0.173 kg equaled 20.2 kcal per day, and that basal energy needs for a human female weighing 56.5 kg were 1,349 kcal per day. If basal metabolic rate increased linearly with body mass, how many kcal would the human female require? Use a calculator to solve this; then imagine the energy needs of a mare weighing 200 kg ("That's a lot of hay.")

Use the Allometric Equation Solver to estimate energy needs for the female rat and human, using these values: a = 70, M = body mass in kg, and b = 3/4 or 0.75. Try it again using b = 1 and compare with your original answer.

What are the practical implications of Professor Kleiber's observation that metabolic needs scale to the 3/4 power of body mass? You can use the **STELLA**® model to compare results for animals of various weights, comparing powers of 0.75 and 1.0.

6.8 Heat and Energy Production Relative to Human Metabolic Needs

The student of biological science seldom remembers the basic units of energy because that subject is usually explained in physics. Considering that obesity is an imbalance between energy intake and energy use, it seems that most of us dozed through our physics classes, and it may behoove us all to learn more about this subject! Most people know about calories in food, but few of us understand the connection between caloric expenditure and force, work, energy, and power. If we are to understand anything about the human body, we must understand the concept of work, so relax and read the following.

Why is work so hard, and why do we try to avoid it? The reason is so commonplace that we usually overlook it. Our bodies are constantly acted on by a force caused by the acceleration exerted by gravity, in an amount equal to our body mass times 9.80 m per sec^2 (close your eyes and feel it). For a person with body mass of 70 kg, the product is a force of 686 kg * m sec^{-2}, defined in units of force called the *newton*. (Imagine an apple accelerating vertically onto the head of a scholar.) So long as we remain at rest, this value does not change. Work is defined as movement of a force through a distance, so our hearts do work to pump blood against the force of gravity, and our muscles perform work when our bodies move in a vertical distance or are used to apply a force to any object we may be carrying. To determine the amount of energy required to do a given amount of work, the force (mass times acceleration) must be multiplied times the distance moved. The work done in lifting a mass of one kg through a distance of one meter is the unit of work or energy called the joule, and the amount of energy required equals (1 kg)(1 meter)(9.807 m sec^{-2}) = 9.807 joules.

Most people are familiar with the unit of heat energy called the calorie, however, the SI system uses the joule. The traditional calorie equals 4.184 joules, and the kilocalorie (kcal) of nutrition therefore equals 4,184 joules. Let us quickly calculate the amount of work required to go upstairs. Suppose that a 70 kg man walked up two flights of stairs with a vertical rise of 10 m; the product is (70 kg)(9.8 m sec^{-2})(10 m) = 6,860 joules. The amount of energy needed therefore equals 1,640 calories or 1.64 kcal—possibly a disappointingly small number for a person interested in weight loss! Solace may be found in the thought that the rate of doing work is discretionary, and the rate of doing work is power—literally!

Let us estimate the energy needed for the heart to pump blood, considering only the effect of gravity and ignoring any effects of friction or viscosity. Assume that our 70 kg person is 1.82 meters in height and that his body is supplied with 5 liters of blood. Assume that the weight of 1 liter of blood equals 1 kg, and that the blood makes one circuit each minute of the day. The work required is therefore (5 kg)(1.82 m)(9.8 m sec^{-2}) = 89 joules per minute. Because there are 1,440 minutes per day, this would require 128,160 joules per day if our test subject remained standing and did no other work. This is only 30.6 kilocalories, a remarkably small value. The actual amount of work is higher due to the fact that the heart muscle must work to pump blood, expand the elastic walls of the aorta, and increase blood pressure against the resistance of innumerable minute blood vessels.

The kinds of work done by the body include mechanical work (lifting heavy objects), electrical work (separating charges across a membrane), osmotic work (concentrating substances against a gradient), and chemical work (biosynthesis and energy-dependent degradation). For now, it is permissible to ignore the thought that electrical current represents a form of work with those dreaded electrical and magnetic units. However, you must remember this: There is a way to convert all forms of work to calories, and calories are easy to convert to joules. Thus food energy is converted to all the other forms of energy continuously so that we may do good works daily, and so that biochemists may ask each new class of students "How many ATPs are produced per glucose molecule in glycolysis and the Krebs cycle?"

6.9 Estimating Daily Energy Expenditure

The first key in determining guidelines for energy intake and diet planning is to estimate daily energy expenditure. Direct measurements of human energy expenditure require room calorimeters, which are very expensive and not available to most individuals. Where feasible, indirect measurements of heat production can be estimated from oxygen consumption, with the assumption that one liter of oxygen is equivalent to about 4.8 kcal of heat production (McArdle et al., 1991). The Food and Nutrition Board uses estimates of 2100 kcal per day for women between the ages of 15–22, 2,000 kcal from 23–50, and 1,800 kcal thereafter (*RDA Handbook*, 10th ed.). For men, corresponding estimates are 2,900 kcal per day for ages 19–22, 2,700 for 23–50, and 2,400 thereafter. Unfortunately, gender- and age-specific estimates may under- or overestimate actual needs by a wide margin. Therefore, it is preferable to use other physical data to improve the prediction of caloric requirements.

Because it is important to provide an accurate estimate of energy needs without using sophisticated equipment, an alternative approach is to employ equations based on factors that influence energy needs. These factors include age, gender, body mass, body surface area, physical activity, the thermic effect of food, climate, and pregnancy or lactation. Several widely used equations are listed in Table 6.1. Although there are many different ways of estimating resting energy expenditure (REE), several recent studies have shown that it is feasible to predict energy needs from body composition divided into fat-free mass and fat mass. In this usage, fat-free mass includes not only skeletal muscle but the brain and all aerobically active, abdominal organs. Nelson et al. (1992) summarized data for oxygen consumption of various organs, which indicate that the brain consumes about 20% of REE, with liver and skeletal muscle each contributing about 25%, and heart and kidneys each adding about 10% of REE. The metabolic rate per kg of adipose tissue is only about 1% of that of the lean tissues, for which reason estimates of REE are not greatly improved by including fat mass.

If one agrees with this approach, one needs a means to estimate body composition, and an equation relating fat-free mass to energy expenditure. The best accepted standard for body composition is still underwater weighing, and electrical impedance also provides an alternative. Sjostrom (1991) recently derived equations for predicting fat mass of adult men and women from studies based on computerized tomography. These were expressed in terms of weight and height, and provided errors of about 7–9%. Therefore, if

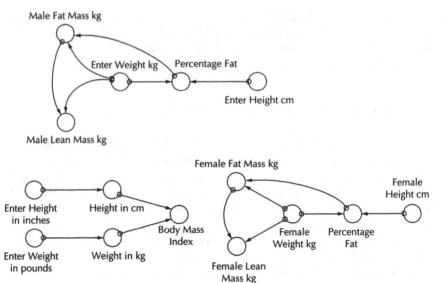

FIGURE 6.3

it is acceptable to estimate REE to within 10%, these equations enable fat mass to be calculated, from which fat-free mass can be obtained by subtraction. Alternatively, girth measurements may be used, which can provide estimates of body composition that are accurate to within 3–5%. This is the approach used by all branches of the U.S. military (Marriott and Grumstrup-Scott, 1992).

STELLA® Application: Create a Model to Calculate Body Composition

Use **STELLA®** to create a model to calculate body composition using the map shown in Figure 6.3 and a set of equations generated by the U.S. Armed Services, as follows:

```
Body_Mass_Index = Weight_in_kg/(Height_in_cm/100)^2
Enter_Height_cm = 178
Enter_Height_inches = 60
Enter_Weight_in_pounds = 160
Enter_Weight_kg = 85
Female_Fat_Mass_kg = (Percentage_Fat *
Female_Weight_kg)/100
Female_Height_cm = 154
```

```
Female_Lean_Mass_kg = Female_Weight_kg -
Female_Fat_Mass_kg
Female_Weight_kg = 160.0
Height_in_cm = 2.54 * Enter_Height_inches
Male_Fat_Mass_kg = (Enter_Weight_kg * Percent_Fat)/100
Male_Lean_Mass_kg = Enter_Weight_kg - Male_Fat_Mass_kg
Percentage_Fat = 0.638 * Female_Weight_kg - 0.409 *
Female_Height_cm + 54.367
Percent_Fat = 0.464 * Enter_Weight_kg-0.411 *
Enter_Height_cm + 54.769
Weight_in_kg = Enter_Weight_in_pounds/2.2
```

STELLA® Application: Calculate Physiological Functions

Use **STELLA®** to solve one or more allometric equations in Table 6.1 for physiological functions for a person of your stature.

TABLE 6.1 Table of Key Relationships in the Human Body

Body surface from height and mass, S (cm^2) = 71.84 * M(kg)$^{0.425}$ * H(cm)$^{0.725}$

Energy expenditure, E = 4.825 * VO$_2$ where E is in kcal and V is liters

Basal metabolic rate = 70.5 M$^{0.73}$ (Benedict)

          or 70 * M$^{0.75}$ (Kleiber, p. 214)

Rate of heat flow across a surface   (Kleiber, p. 148)

$$q = A * (T_b - T_a)/r$$

where $r = L/\lambda$

     q = rate of heat flow per unit time

     A = cross-sectional area of flow

     T_b = body temperatute

     T_a = ambient temperature

     r = specific insulation where L = thickness and λ = heat conductivity

Heat conductance $\lambda = \lambda(A/L)$

Pulse rate (bpm) = 186 * M(kg)$^{-0.25}$ (Calder, p. 192)

Blood volume = 65.5 M$^{1.02}$ (Calder, p. 145)

         or 0.069 M$^{1.02}$ (Total blood volume is 7% of mass; plasma volume is 4%)

Fat mass = 0.075M$^{1.19}$

Fat mass, males (kg) = 0.923(1.36 * W (kg)/H (m) − 42)   (Sjostrom, 1991)

Fat mass, females (kg) = 0.923 (1.61 * W (kg) /H (m) − 38.3)

Muscle mass = 0.468 M$^{1.0}$

Turnover times of metabolic pools, t_t = P/r = k_1M/k_2M$^{0.75}$ = kM$^{0.25}$ (Kleiber, p. 216)

Life span = 11.6M$^{0.20}$ (Calder, p. 153)

Mortality doubling time t_2q_0 = 0.156 M$^{0.27}$ (Calder, p. 315)

Survivorship $n = n_0 e^{(q_0/\alpha)1-e^{\alpha t}}$

     where q_0 =0.0124 * M$^{-0.56}$ and α = 0.709 M$^{-0.27}$ (Gompertz equation, Calder,
                                          p. 315-316)

References

Benedict, F.G. *Vital Energetics: A Study in Comparative Basal Metabolism*. Washington, D.C.: Carnegie Institute of Washington Publication #503, 1938.

Calder, W.A., III. *Size, Function, and Life History*. Cambridge, Mass.: Harvard University Press, 1984. See Table 3.4, Allometric comparison of the internal proportions of body mass in mammals and birds, p. 48–49.

DuBois, D., and E.F. DuBois. "Clinical calorimetry. A formula to estimate surface area if height and weight be known." *Arch. Internal Med.* 7, 863–871 (1916).

Durnin, J.V.G.A., and J. Womersley. "Body fat assessed from total body density and its estimation from skinfold thickness: Measurements of 481 men and women aged from 16 to 72 years." *Br. J. Nutr.* 32, 77–97 (1974).

Ellis, K.J. "Reference man and woman more fully characterized. Variations on the basis of body size, age, sex, and race." *Biol. Trace Element Res.* (1990): 26–27, 385–400.

Hodgson, J.A. "Body composition in the military services: standards and methods." In Marriott, B.M. and J. Grumstrup-Scott, eds. *Body Composition and Physical Performance. Applications for Military Service*. Washington, D.C.: National Academy Press, 1992, 57–70.

ICRP, "Report of the Task Group on Reference Man: Anatomical, Physiological, and Metabolic Characteristics," *ICRP Report No. 23*, Oxford, England: Pergamon Press, 1975.

Kleiber, M. *The Fire of Life. An Introduction to Animal Energetics*. Robert E. Krieger Publishing Co., Inc., Huntington, N.Y., 1975.

Lentner, C., C. Lentner, and A. Wink. (1986) *Geigy Scientific Tables*, 8th ed. West Caldwell, N.J.: Medical Education Division, Ciba-Geigy Corporation, 217–231.

Marriott, B.M., and J. Grumstrup-Scott, eds. *Body Composition and Physical Performance. Applications for the Military Services*. Washington, D.C.: National Academy Press, 1992.

National Research Council. *Recommended Dietary Allowances*. 10th ed. Washington, D.C.: National Academy Press, 1989.

Nelson, K.M., R.L. Weinseir, C.J. Long, and Y. Schutz, "Prediction of resting energy expenditure from fat-free mass and fat mass." *Amer. J. Clin. Nutr.* 56 (1992): 848–56.

Sjostrom, L. "A computer-tomography based multicompartment body composition technique and anthropometric predictions of lean body mass, total and subcutaneous adipose tissue." *Int. J. Obesity* 15 (1991): 19–30.

Thomas, P. R. *Weighing the Options. Criteria for Evaluating Weight Management Programs*. Washington, D.C.: National Academy Press, 1995.

7

A Primer on Biodynamics and Gene Expression

A rather small number of relatively simple structures will be found repeatedly . . .

Jay W. Forrester

The most crucial element in scientific research is *hypothesis testing*. An hypothesis may be defined as "a tentative assumption made in order to draw out and test its logical or empirical consequences." An hypothesis must be stated in reference to a specific *conceptual model* that can be demonstrated to be false, if that is the outcome of a test. Before scientists ever begin an experiment, they review current information that pertains to a system of interest, and ask whether important gaps in knowledge exist. In this process, they are creating a mental model of a system, which is not merely an abstraction. To a new investigator, choosing a research topic in a highly competitive research environment is the most important career decision imaginable. The importance of simulation relative to this process is that the models can be created as system diagrams that actually function to produce numerical predictions concerning behavior. Even though the data are in a sense imaginary, there are practical applications for this ability. I have received at least one federal grant in which simulation was used in the grant application to explain an important concept to the review panel. The technique can also be used to generate system diagrams and data for seminars.

The branch of mathematics that was created to deal with movement and rates of change is the calculus, which, to most biologists, is memorable for how little of it can be remembered when the final examination is over. Most important is not whether the student ever understood differential calculus, but whether one ever uses it to formulate a problem. Admittedly, there exist fields of study in which differential equations are requisite, but this was not true of any laboratory I ever worked in. Our conceptual diagrams often indicated ways that molecules could accumulate or disappear over time, but even though we wished to *integrate* information, we never *differentiated* in the mathematical sense. Biologists are very accustomed to integration, but do not understand why it should be necessary to consider infinitely small units of time to arrive at an answer. The fundamental process used in the calculus seems *unnatural* to biologists; in the words of Jay Forrester, "Nature does not differentiate; she integrates." If one divides up DNA into ever smaller parts, eventually one finds that the substance is made up of nucleotides, and there is no need to continue subdividing ad infinitum. Dividing up proteins, one finds amino acids, and few biologically interesting events occur at lower levels of structure. So it is that biologists tend to think in terms of units or quanta, and it makes sense to examine interactions and transfers among these quanta. For this to be successful, one needs to be able to add and subtract, sometimes multiply and divide, and even to use equations that define non-linear responses. However, one does not need to differentiate!

This chapter compares simulation and modeling as methods of approaching scientific questions, and introduces a method called *compartmental analysis*, which is applied in the context of gene expression. Although there are numerous methods by which experimental data can be analyzed, compartmental analysis is a simple and natural method that can be applied to understanding system dynamics in general, and biodynamics in particular. This method produces outcomes that are simulations, or predictions based on changes in specific variables that influence a system. Modeling requires comparison between outcomes of simulations and observational data, or between theory and experimental results, which is part of a process by which understanding may be improved.

7.1 Simulation and Modeling Can Assist Analysis and Integration

The capacity to forecast future outcomes of current decisions and actions is uniquely human. It is an activity that must be mastered not only by scientists, economists, and health care practitioners, but also by every parent who ever hopes to raise a child from infancy to adolescence. It is a key element in applying for funding to carry out medical research, for review teams are very interested in assessing whether or not a proposal is likely to yield significant new knowledge. The need to forecast quantitatively is a feature that is strongly emphasized in medical science; it is important to learn to state ideas in terms of hypotheses, and equally important to explain means to quantify results, so that the likelihood that an hypothesis is true or false can be stated using numbers and statistics. In reading and evaluating scientific literature, the ability to understand and extrapolate from data contained in tables and graphs is fundamental, every bit as much as the need to practice shooting, dribbling, and passing is fundamental to skill in basketball. The student who understands that numbers represent a uniquely human way of interpreting the mysteries of the natural world may have the qualities needed to become a scientist, for scientific and medical thinking demands the ability to think in this somewhat unnatural manner. Whether or not every person has this capacity may be unclear, but those who have come to a new discovery or prediction using this set of tools often consider it a peak experience that can be as moving as infatuation between a man and a woman.

Scientific theory depends on being able to predict outcomes over time, and this involves several key components. The first key is the ability to identify and measure each essential factor that controls the behavior of the system of interest. For instance, the theory that intake of cholesterol and saturated fats contributes to coronary vascular disease required a means to measure cholesterol and other lipids in the diet, in blood plasma, and in atherosclerotic plaque. Once the association became clear through epidemiological and pathological studies, the question of what controlled cholesterol levels in the bloodstream arose. This in turn required means to measure cholesterol in lipoprotein particles, enzymes involved in biosynthesis and metabolism of cholesterol, and receptors involved in cholesterol recycling. With the advent of the ability to quantify messenger RNA and to isolate genes, the tools advanced to a still more fundamental level. In the course of learning how to measure system components, medical scientists perform experiments in a second key area, namely, how the components change over time if one variable item is changed while others are held constant. Now a dynamic or kinetic aspect has been introduced into the experiment, and for a theory to fit with experimental results, it is important to be able to ask whether the observed rates of change would have been predicted on the basis of the known parts of the system. If not, that system component may not govern the response, or there may be unknown features that must be identified and explored.

Biomedical research progresses from an analytical period, which involves identifying the functioning parts of biological systems, to an integrative period in which experiments test for important system interactions. The researcher usually begins with an effect, such as the observation that patients who suffer from heart disease usually have blockage of the coronary arteries. He or she then works backward toward an understanding of the causative agents. The beginning student may tend to seek causal chains, but mastery requires that one come to grips with multiple factors that interact in a webwork. And it is a web, for every human tissue contains numerous cell types, dozens of signals, hundreds of components, and thousands of active genes and gene products. Given this complexity, it is not difficult to feel buried, and thus desirous of a tool that can assist in dealing with complexity.

7.2 Simulate to Think, But Model to Validate

Simulation, in the purest sense, refers to methods of observing hypothetical outcomes without necessarily referring to experimental data. Modeling implies the opposite; one generally has a data set, and would like to identify one or more mathematical models that produce a theoretical curve which fits the data with high accuracy. The highest aim of the scientist is to understand nature, an endeavor that can not succeed without measurement, quantification, and the formation of mental models. In some disciplines, it is feasible to derive equations that predict system properties that were unknown prior to the application of this unique form of logic. Albert Einstein once wrote: "The scientist finds his reward in what Henri Poincaré calls the joy of comprehension, and not in the possibilities of application to which any discovery may lead." Using new software for modeling will not convert a bench scientist into a pure theoretician, but it will allow a mode of thinking that is impossible without this quantitative tool, and proper use of the tool indeed can produce revelations.

One should never underestimate the power of a simple tool as an aid to the advancement of theory. Among the best examples of this are two discoveries in what became chaos theory (Gleick, 1987). In one case, Robert May, using only a hand-held calculator to solve a well-known logistic equation, demonstrated that populations of organisms could show chaotic behavior. In the second example, Edward Lorenz used computer simulation to find that small changes in initial conditions could greatly influence forecasts of future weather patterns in what came to be called the Butterfly Effect. These outcomes of simulation have had great influence on research scientists, and the same tools are available to anyone with the initiative to master the software.

7.3 Compartmental Models

To most English-speaking people, the term "compartment" means an enclosed physical space. The biologist would immediately assume that, in reference to living organisms, a compartment is a physical space such as the blood volume, the extracellular fluid, or the contents of an organ. Even the word "cell" originally was an analogy between microscopic structures and the rooms in which monks dwelled or the components of a honeycomb—it defined a minute physical space. Although there is no reason to change this definition, in modeling and simulation the term usually refers to the *contents* of an imaginary space, rather than the space itself. For example, in a single-compartmental model of messenger RNA production, the item of interest is the mRNA, and the space in which it is contained is mathematically irrelevant. However, a researcher interested in the precursor to the mature mRNA might formulate a two-compartment model in which the contents of the nucleus moved into the cytoplasm. All items of interest to biologists are present in some kind of physical space, and it is a peculiarity of modeling that the quantity of material and the volume of the space it occupies must be indicated separately.

A compartment may be defined as any kind of unique, homogeneous, and identifiable material of interest to an investigator. Examples include a specific drug that is present in the blood plasma, a specific mRNA or protein, or a specific population of cells. The material may exchange with other compartments; the drug may be metabolized in the liver to form a distinct but identifiable metabolite, or the protein may be phosphorylated and converted to an alternate form with distinct properties that could be measured immunologically or by a chemical assay. In the Stock-and-Flow notation used in the **STELLA**® program, a stock represents an accumulation of one kind of material; its *contents* are a compartment and the icon itself is only a symbol.

Compartmental analysis also represents a mathematical technique to describe the curve that represents the change in amount of substance over a period of time. Compartmental analysis is a method of quantifying the rates of accumulation, transfer, and elimination of material or information in systems with one or more uniform physical or chemical states (Godfrey, 1983; Jacquez, 1985; Green and Green, 1990). For example, a drug in the bloodstream might be eliminated by the kidneys, which would create a rate of elimination. Movement into the liver would create a second rate of disappearance, and metabolism in the lungs could represent a third rate. Compartmental modeling describes each of these rates of disappearance as separate *kinetic compartments*, which really means that each compartment represents a different form of the substance, each described by a different exponential function. Investigators who attempt to monitor the metabolism of nutrients and drugs by means of radioisotopic tracers are frequently heard to say, "We started with a three-compartment model and wound up with a nine-compartment model," meaning that an equation with nine separate exponential terms described the data well. I once heard a presentation concluding that the best fit was obtained with a 27-compartment model. The answer to the question of what physical space or processes corresponded to the 27-compartments was "None of which we are aware!" To which a feeble voice responded, "How about the liver?"

7.4 A Compartmental Model of Gene Expression

The flow of genetic information through the steps of transcription, mRNA processing and transport, and protein synthesis and degradation is a familiar example of linked processes that are often diagrammed qualitatively. Each gene product presumably contributes to a specific trait or phenotype, and is influenced by signals and stimuli from a network of other cells, tissues, or the external environment. This diagram can be used as a framework for a quantitative, mathematical model by assigning rates to each connection indicated in the diagram; common ways to represent the rate parameters are with the English letter, k, or with the Greek letter, lambda (λ), with subscripts identifying which rate applies to each specific process. The rates need not be constants, as long as the modeler states how a change in input is related to a change in output, or how a stimulus is related to a response. This relationship can be as simple as the fraction of the component of interest that is transferred during each interval, an algebraic equation, or a nonlinear saturation function such as the Michaelis-Menten or Hill equation.

Compartmental modeling is ideally suited to studies of the flow of genetic information, in which latent information is transcribed, amplified, and translated to generate an enzyme capable of altering the properties of a system.

The material of interest is defined by the observer; it may be a new mRNA transcript, a specific protein, total cellular mRNA, or the nucleotides or aminoacyl tRNAs used to support the biosynthesis of these molecules. Compartmental analysis functions best in quantitative studies of molecules within a cell *population*. Treating protein synthesis as though mRNA or protein in a single cell were under study has little biological relevance, because the typical organ contains approximately 10^8 cells per gram, and samples required for most biochemical analysis represent thousands of cells. The behavior of a single cell is irrelevant to most mammalian traits; what is of interest is how the combined gene products in a cell population contribute to the economy of the organism. In the absence of entraining agents such as hormones, mitotic signals, or altered membrane potential, any oscillations within individual cells would be masked. In this view, a phenotype is a statistical trait, and the investigator recognizes that an experiment is an exercise in estimating a population mean through statistical methods. Compartmental analysis seeks to estimate parameters that describe the dynamics of identifiable populations or "pools" that show uniform behavior.

7.5 Kinetic Behavior of a Single Compartment

The single-compartment, quantitative model can be viewed as a kinetic unit of function. It refers to a product that is formed at a fixed rate under initial conditions, and that is eliminated at a rate proportional to concentration; these are defined as zero-order filling and first-order decay, and the process may be depicted as in Figure 7.1.

In this model, a molecule has a concentration at any moment designated C_t, and is being produced at a rate of synthesis defined as λ_1, expressed as molecules produced per unit of time and tissue mass. The rate of decay, however, is proportional to concentration and is calculated by multiplying the concentration times the first-order rate constant for decay, λ_2, which has units of time^{-1}. *Note: Throughout this book, the symbols λ and κ will be used interchangeably to signify first-order rate constants.*

This quantitative model was introduced in the context of enzyme induction by Monod, Pappenheimer, and Cohen-Bazire (1952) during studies of the effects of lactose analogs on β-galactosidase in *Escherichia coli*. The studies assumed that synthesis of this enzyme initially occurred at a constant rate (zero-order kinetics) that could be increased in the presence of substrate when glucose was limiting for growth, and that enzyme decay was proportional to concentration (first order). When the assumptions of zero-order synthesis and first-order decay are valid, product will accumulate according to a simple, exponential function that can be expressed as follows:

$$C = \frac{\lambda_1}{\lambda_2}(1 - e^{-\lambda_2 t}) \qquad (7.1)$$

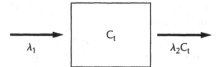

FIGURE 7.1

This equation predicts that, at equilibrium, the concentration (C) of β-galactosidase equals the ratio between its rate of synthesis (λ_1) and the rate constant for decay (λ_2). In this example, the rate of synthesis had units of enzyme activity produced per milligram of bacterial protein per unit of time. The rate constant for decay equals the fraction of the existing enzyme that is removed during each interval of time, and has units of time^{-1}; dividing the rate of enzyme synthesis by the fractional rate of degradation gives units of enzyme per mg of protein. If one of the kinetic parameters changes, the concentration will approach a new steady state in an exponential fashion, with a period defined by the rate constant for decay. The exponential term, $1 - e^{-\lambda_2 t}$, is a fraction that varies between the values of zero and one.

The quantitative model used by Monod and colleagues is an example of a single building block of the technique called compartmental analysis. In this case, the single "compartment" was all the enzymatically active β-galactosidase in the cell culture. The form of systems theory that employs compartmental analysis simply establishes the flow of material into and out of each distinct, kinetic compartment of interest to the experimentalist, and produces a set of equations that describes the entire set of interactions. The technique is applicable in molecular biology because the same kinetic model was later used to analyze changes in concentrations of enzymes and mRNAs in mammals, birds, and insects (Berlin and Schimke, 1965; Hoel, 1970; Hargrove, 1993). When the relationship between mRNA production and protein accumulation is examined, a two-compartment model is produced. This idea may be readily extended to include nuclear and cytoplasmic processes, as well as the external stimuli that alter gene expression in the system.

The idea that decay constants figure importantly in determining the period required for enzyme induction was recognized by Berlin and Schimke (1965) during comparative studies of enzymes in mammalian liver. They concluded that the period required for several enzymes to shift to new levels after hormonal stimulation was related to their half-lives, even when rates of synthesis were similar. According to this kinetic model, *the decay constant is equivalent to the time constant for change in product concentration*, because the period required to achieve half of the total shift in concentration between two steady-state levels cannot be less than the half-life of the mRNA or protein under study. This idea may be restated in terms of compartmental analysis very simply by saying that the chemical half-life is one measure of the fractional transfer rate from the compartment of active mRNA or protein to the "compartment of oblivion." Because the half-lives of these molecules vary from a few minutes to several days, the period required to achieve a new steady state varies within the same range.

The models for interpreting mRNA and protein induction as a function of time are not difficult to unify in principle. In the simplest case, one assumes that the rate of protein synthesis is proportional to mRNA concentration, and seeks to quantify the number of protein molecules synthesized per mRNA per unit of time. When this is done, one has begun with what is technically called a single-compartment model, and created a dual-compartment, mathematical model of gene expression. The fact that gene expression could be modeled by the technique of *compartmental analysis* was recognized by David Hoel (1970). Theorists recognize this as one of the most common general techniques of systems theory, and basically the same model was used by other mathematicians (Goodwin, 1963; Maynard-Smith, 1968; Yagil, 1980) before and after Hoel's analysis.

This book will follow this precedent for two reasons. First, compartmental analysis is extremely straightforward and has been used by nearly all experimental scientists who have attempted to analyze gene expression quantitatively. Second, some of the simplest, but most powerful, microcomputer programs available for mathematical modeling are based on this method and can be used appropriately by most biologists. Readers who learn this method are advised to always justify the underlying assumptions on which their models are based, and to locate collaborators who are trained in advanced methods of numerical analysis if unusual behavior is encountered.

7.6 Assumptions and Nomenclature Used in Modeling Systems

One important reason for modeling is to formulate and test hypotheses by determining whether observed experimental results fit with predictions based on current understanding of how a biological system operates. The modeler typically compares the highest quality experimental data for the system of interest to the response curves predicted from the best available quantitative theory. Inevitably, discrepancies are observed that provide evidence for unsuspected interactions, ways to improve the theory, or both, and the modeler seeks to improve his or her understanding of the system by improving the experimental techniques and the mathematical theory. Many current computer programs also provide statistical techniques for estimating parameters that provide the best fit between a conceptual model and the measured data.

Systems theory includes models that may be either deterministic or stochastic. *Deterministic behavior* implies that the future state of a system is predictable based on knowledge of its current state and any future inputs, whereas *stochastic behavior* includes an element of randomness, heterogeneity, or uncertainty about the future state of the system. The idea that a tenfold increase in mRNA concentration will produce a tenfold increase in the concentration of the corresponding protein is an example of a deterministic model that is also *linear* because a change in the concentration of mRNA produces an equivalent change in protein when the system attains equilibrium. Behavior is linear if a change in input produces a proportional change in output. Although the assumption of linear behavior is seldom questioned, most biological processes are nonlinear if examined over an appropriate dynamic range. For example, enzymes and transport systems become saturated as substrate concentration increases, and this produces a *nonlinear response curve* typified by hyperbolic, Michaelis-Menten kinetics. If a tenfold increase in mRNA concentration only produced a sixfold increase in protein, then the system might be behaving nonlinearly, or might not be in equilibrium. The reason that compartmental analysis works very well for analyzing genetic systems is partly attributable to the fact that an increase in transcription often produces a proportional increase in mRNA concentration followed by a proportional increase in protein synthesis. If one considers the complexity of the process, this is surprising, but cell-free translation of mRNA has shown that responses are linear across a moderate range. In one of the original systems for cell-free protein synthesis, Pelham and Jackson (1976) showed that globin mRNA was translated in proportion to concentration at ranges up to 10 μg/ml, and Palmiter (1973) obtained a linear response for ovalbumin synthesis at mRNA concentrations as high as 200 μg/ml. The response to mRNA in vivo is more difficult to ascertain, but Richter and Smith (1981) observed a linear increase in globin synthesis in oocytes from *Xenopus laevis* when quantities of 5–100 ng of mRNA were injected per egg.

Most biologists assume that a system of interest can be manipulated to produce an initial *steady state*, which is defined as a condition in which the material of interest is produced at the same rate as it is degraded or eliminated. When this is true, the concentration of material will be constant, or time-invariant. For example, cells may be grown in a defined tissue culture medium until they reach confluence, so that a particular mRNA is maintained at a fixed concentration. The initial conditions are then changed by the addition of a hormone, cytokine, or nutrient, and the mRNA concentration is measured as a function of time. Although true steady states do not occur in biological systems, the assumption facilitates model-based analysis and is implicit in most of the literature in molecular biology. After the system has been perturbed, the mRNA (or other dependent variable) is assumed to approach a new steady state that depends on the concentration of the inducing stimulus.

7.7 Transfer Rates Between Compartments Are Related to Half-Lives

Visualize a system of gene expression in which an investigator had treated cells with a specific inducing agent, and was attempting to relate the accumulation of mRNA in the cytoplasm to the rates of transcription; nuclear processing, decay, and transport; and cytoplasmic degradation. It should be self-evident that the rate of transcription responds first, and could come to a new equilibrium before the nuclear transcript attained its new steady-state level; likewise, the concentration of mature mRNA in the cytoplasm could only reach an equilibrium after the preceding processes had attained new, constant rates. This "thought experiment" exemplifies the most basic tenet of compartmental analysis: *rates of transfer between compartments determine the time required to approach a new steady state*. Note that it is not necessary to actually achieve the steady state; the model suggests that transfer rates determine the rate of approach to steady state whether or not there is time to attain the new value.

Observe also that (1) rates of degradation or metabolism are equivalent to rates of transfer, transport, processing, or secretion in compartmental analysis; and (2) these rates combine to determine the time course of response for the system. In compartmental analysis, individual compartments are diagrammed as circles or rectangles with arrows connecting each one to the next compartment or to the rest of the system. Each transfer function has the units of reciprocal time, or amount of material transferred per unit of time. In other words, each arrow represents a rate coefficient, or fractional transfer rate. *Fractional transfer rates are not kinetic constants*; they simply indicate how much of the material in the pool moves out of the pool by the indicated route during any interval. Usually, these rates are treated as constants, but this is not essential to compartmental analysis. If the investigator determines that the transfer rate varies as a function of concentration or time after a stimulus, then it is very easy to modify the program to indicate the correct relationship with a hyperbolic curve or sliding function.

As the fraction of material transferred or eliminated from a compartment per time interval increases, the shorter the half-life becomes, and the quicker the system approaches its new equilibrium. Therefore, the fractional transfer rate (or fractional elimination rate, λ) is inversely proportional to the half-life ($t_{1/2}$). By definition, the relationship is stated as follows:

$$\lambda = \frac{0.693}{t_{1/2}} \tag{7.2}$$

The meaning of this equation is that material that is transferred or degraded rapidly has a short half-life. In other words, half-life is a derived property that is related to the stability or persistence of a mRNA or protein in the cell, presumably because each protein and mRNA interact with cellular systems for degradation or transport to differing degrees.

7.8 Compartmental Models Help Predict Steady-State Relationships

An important attribute of compartmental modeling is that it permits the investigator to build a system using a flow chart or set of block diagrams. After rates have been assigned, the steady-state relationships may be written with little practice, and can be solved using a hand-held calculator. No knowledge of differential equations is necessary to do this, because in the initial or final steady state there is no change. This is a very useful method for individuals who think better within a visual context than within a numerical context. The diagram is constructed on the computer screen using the icon-based tools provided by the simulation software, and numerical values or equations are typed in at the keyboard by means of dialog boxes. This method overcomes fear of mathematics, because the investigator is working visually and the computer itself is fearless. All the investigator needs to provide is a diagram linking the components under investigation, the rates of transfer, and a notion about whether the components are related in a linear or nonlinear manner.

7.9 Simulating Gene Expression Using a Single-Compartment Model

This exercise introduces the student to principles that represent the foundation of biological dynamics, and explains how to use a microcomputer to understand time-dependent and quantitative aspects of cause and effect. When the exercise is completed, the student should be able to draw a simple diagram of a biological system, convert that diagram to a steady-state equation, and use the microcomputer to solve the equation. The examples used here will be drawn from studies of nutritional effects on genetic regulation.

Zero-Order Synthesis and First-Order Elimination or Decay

The simplest kinetic model and the one which probably has the greatest number of applications involves material that is being generated at a fixed rate, but that moves out of the system or is degraded at a rate proportional to the amount in the system. Let us use **STELLA®** to create and analyze this model. Figure 7.2 shows the model-building interface for **STELLA®**. Access this screen by opening the program and moving the cursor to the down arrow on the left side of the screen. When that arrow is clicked once by pressing the mouse button, the frame shown in the figure will appear.

To construct this diagram, move the cursor to the rectangular icon (called a *stock*) in the top left margin and click the mouse button. Then move the cursor to the middle of the screen and click the button again. This deposits on the screen an icon that represents a single kinetic compartment. Name it by typing Quantity. Now add inflows and outflows by clicking on the next icon, which is called a *flow* (it is a pipe with a valve). Place the cursor to the left of the stock, click the button and hold it down, and drag the arrow into the stock icon. When the stock box becomes gray, release the mouse button. Name the flow by typing it in when the icon is darkened (you may select any icon at any time by clicking on it).

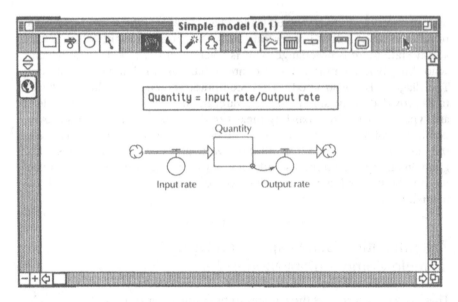

FIGURE 7.2

Repeat this process to add an outflow as shown (If you press the option button on the keyboard, the cursor will become the last tool that was used, so you do not have to return to the title bar to get the tool.) This time, click-and-drag starting within the box, and move out of the stock box before releasing the button. To indicate that decay is proportional, use the fourth tool on the title bar, which looks like an arrow. This is the *connector*. Click within the stock icon, and drag the cursor out to the circular element on the outflow. Release the button when the flow becomes gray. Voila, you have created a diagram that indicates constant input and proportional output. It could represent synthesis and degradation of mRNA, glucose uptake and use by a tissue, or inflow of cash into your checking account and your monthly expenditures. To paraphrase the folk singer, Arlo Guthrie, "you can get anything you want at **STELLA®**'s restaurant."

Use Your Diagram to Create a Computer Program for Kinetic Analysis

Your diagram was created in the mapping mode, as indicated by the map of the planet in the left margin. Now switch to the equations mode by clicking the planet icon. The diagram will look like Figure 7.3.

To convert this diagram into a computer program, one must now enter rates and values in the appropriate places. This is done by opening dialog boxes associated with each icon containing a question mark; click twice rapidly on the stock labeled Quantity. Figure 7.4 will appear on the screen.

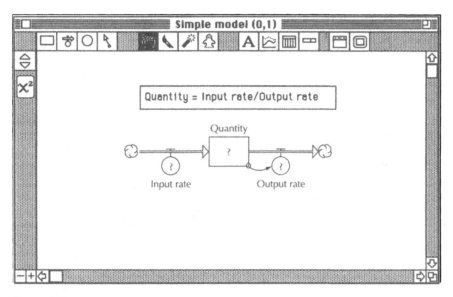

FIGURE 7.3

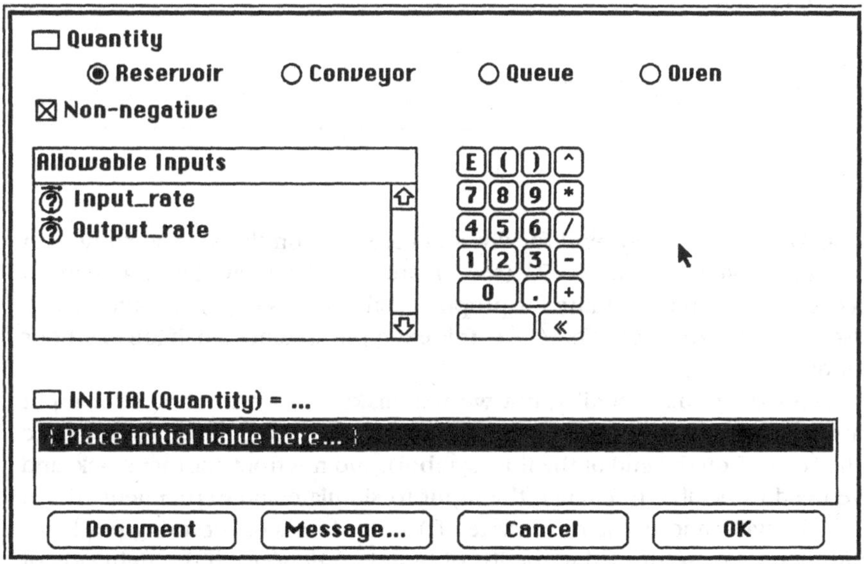

FIGURE 7.4

The highlighted area requests an initial value, which may be a number such as zero or any other quantity, or may reflect the input or output rates. If the output rate parameter is known, a universal value can be entered based on the steady-state equation, Quantity equals Input Rate divided by Output Rate Parameter. Click the term Input Rate once, and it will enter the highlighted

◻ INITIAL(Quantity) = ...

Input_rate/0.693|

FIGURE 7.5

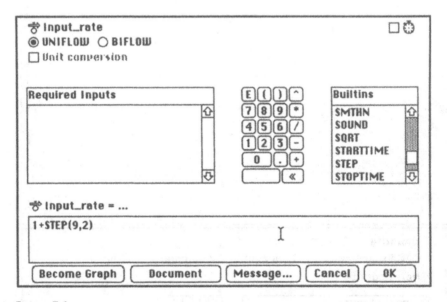

FIGURE 7.6

area. Then click the symbol for division (or slash) on the keypad shown on the screen, followed by 0.693 (see Figure 7.5). This number was chosen because the elimination rate for proportional decay is 0.693/half-life, which also equals ln2/half-life. Therefore, this example assumes a half-life of 1 hour (or other time unit).

Each dialog box is similar, but we will make a different choice for input rate. Our diagram indicates that input comes from an inexhaustible source (the "cloud" on the end of the flow symbol), and not from another stock, and we need to be able to change the input to simulate and experiment. To do this, we will choose a starting value of 1 unit of material entering per hour, and then increase this using the list of built-in functions on the right side of the screen in the dialog box for the input rate, shown in Figure 7.6. First, type in the value, 1, followed by a + sign. Then scroll down the list using the arrow on the right side of the list, or by dragging the open box downward until you see the Step function (a term that increases a rate by a specified increment at a specific time). Click it once, and it will enter into the highlighted area, with the cursor ready to insert values. If you click on the button labeled OK at the bottom of the screen, a message will tell you what kind of information to type in. Here, we need the height and timing of the change; (9, 2) means that an increase of 9 occurs at 2 hours, as shown, so the final input rate is 10 units per hour.

Similarly, open the outflow icon and type 0.693*Quantity to indicate the outflow rate we specified. At this point, we have a completed computer program ready to run. Verify this by checking the equations you have built. Click the arrow above the X^2 button, and the equations should be displayed as follows:

```
Quantity(t) = Quantity(t - dt) + (Input_rate -
Output_rate) * dt
INIT Quantity = Input_rate/0.693

INFLOWS:
Input_rate = 1 + STEP(9,2)
OUTFLOWS:
Output_rate = Quantity * 0.693
```

7.10 Writing Equations for the Steady-State Condition

It is possible to calculate the initial and final values for the material of interest very easily, which is useful in setting up the computer program. If the wrong value is chosen, the system will first try to find its equilibrium position, so that the initial curve will be either increasing or decreasing in an undesirable way. Here is how to avoid this problem.

Examine the basic system diagram and ask yourself what the input and output rates are. At equilibrium, there is no change in the values (steady state), so the rate of change is zero. Therefore, input equals output. In the simple model shown, this means that we can represent the rates as shown:

Input rate (k_1) equals Quantity times Output rate (k_2),

or

$$k_1 = k_2 Q$$

Therefore,

$$Q = k_1/k_2$$

Now ask yourself what would happen if k_1 or k_2 changed to a new value. The system would come to a new steady state, but the same equation would hold; the only difference is that there would be a new value for one or both of the constants.

To make the program work correctly, there is only one other item you need to know: the relationship between the time required for the system to change by half (biological half-life) and the rate of elimination, k_2. The rate of elimination is also known as the fractional decay rate. It specifies what fraction or percentage of the material is removed from the system during the period of interest, and the period may be in minutes, hours, days, or any other time frame. For genetic changes, the time frame is usually expressed in terms of hours.

Example

The rate of synthesis of a mRNA is 1 molecule per hour per cell, and the mRNA half-life is 1 hour. What is the concentration? First calculate the fractional decay rate:

$$k_2 = 0.693/1 \text{ h} = 0.693 \text{ h}^{-1}$$

Substitute this value into the steady state equation:

Q = 1 molecule per hour/0.693 per hour = 1.4 molecules per cell

Now assume that the rate of transcription increases ten-fold. When the system comes into equilibrium, what will be the new concentration?

New concentration = _____

Now suppose the mRNA is stabilized so its new half-life is 4 hours. Calculate the new concentration. _____

Summary

You can write an equation that applies to any system by writing the equation for each component, as you just did for an mRNA, and then adding all the equations together. This will give you initial and final values for the computer program. The rules are as follows:

1. If a rate does not depend on how much material was in an earlier compartment, then it is just a simple number, such as the 1 molecule per hour per cell from the example.
2. If the rate does depend on how much material is in a preceding compartment, then the rate of input equals a constant times the material in the compartment. In our example, the rate of outflow was 0.693 times the quantity of mRNA.

   Note: In the model shown, when a connector is used to indicate proportionality, the user should equate this with a rate parameter, k, with units of time^{-1}. Therefore, in writing an equation based on a diagram, when you see a connector, be sure the rate term equals k_i times quantity in the preceding stock.
3. At equilibrium, the concentration in any compartment equals the sum of all the input rates divided by the sum of all the output rates.
4. In general, the rate of change in the system depends on the sum of all the output rate constants in effect at the time of interest. We will now use the computer program to test these ideas and put them to use.

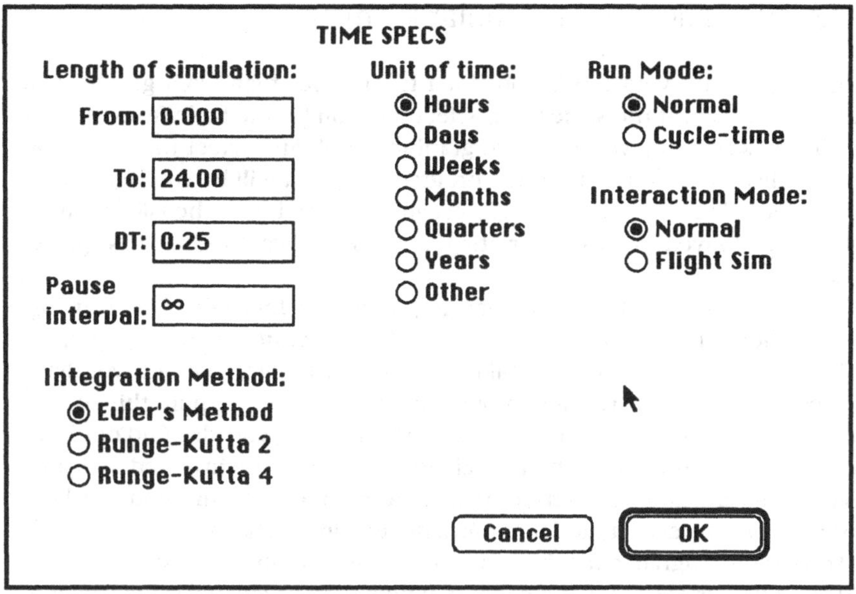

TIME SPECS

Length of simulation:	Unit of time:	Run Mode:

From: 0.000

To: 24.00

DT: 0.25

Pause
interval: ∞

Unit of time:
- ◉ Hours
- ○ Days
- ○ Weeks
- ○ Months
- ○ Quarters
- ○ Years
- ○ Other

Run Mode:
- ◉ Normal
- ○ Cycle-time

Interaction Mode:
- ◉ Normal
- ○ Flight Sim

Integration Method:
- ◉ Euler's Method
- ○ Runge-Kutta 2
- ○ Runge-Kutta 4

[Cancel]   [OK]

FIGURE 7.7

7.11 Test How the Basic Kinetic Model Predicts Change over Time

Now run the program to find out how the model will change when the rate of input changes. From the Menu bar, select Time Specs and change the screen to look like Figure 7.7. This indicates that we will leave the calculation interval at 0.25 h, but change the total period from 12 h to 24 h, and indicate that our time units are hours by clicking the radio button next to Hours. We're ready to run! The output can be viewed as a table, a graph, or both.

7.12 Viewing Results as Graphs and Tables

To see the output, click the graph icon from the title bar, and drag it onto the diagram. Click it twice to open the box that permits one to select the items to be plotted. Highlight Quantity and press the double right arrow in the middle of the screen to enter it; then click OK. Select Run from the Run menu, and watch what happens.

In the initial trial, the scale for concentration arbitrarily began at a value of 1.4. To change it to zero, go to the Run menu and open the item labeled Range specs. Highlight the variable, Quantity, and then type in 0 for the scale, Min(imum). Press the Set button underneath the box to accept the number, and then press the OK button at the bottom right corner of the screen.

7.13 Advantage of the Tabular Form

More precise answers can be obtained from tables than from graphs, and both can be used at the same time. Select the icon for the table from the border in the same way you used the graphical tool, and select the items you wish to display. When you run the program, the data will be entered as a list of specific values. The table and graph can be printed from the File menu, or selected and saved as files for printing in other applications, such as spreadsheets or graphics packages.

Figure 7.8 shows a dialog box for a graph; the one for a table is similar. Figure 7.9 shows a run menu item that is used for resetting the scale of graphs.

Your graph should now look like Figure 7.10. Congratulations, you have successfully built and run a computer model. Now let's apply this tool to understand the basis for change in biological systems, and also find out if we can introduce more typical biological rates, such as those involved in ligand-receptor interactions and enzyme-substrate interactions, and find out how this all fits together in a model of nutrient-gene interactions.

Look at the diagram and table, and answer two questions. First, how much did the mRNA increase? Was it consistent with your calculated value? Second, at what time did it attain half of the maximum response? How does this compare with the half-life you used in the program? Finally, how much time passed before it reached equilibrium? Did this agree with the rule that five half-lives would pass before a new steady state is approximated (actually, 98% of the new value)?

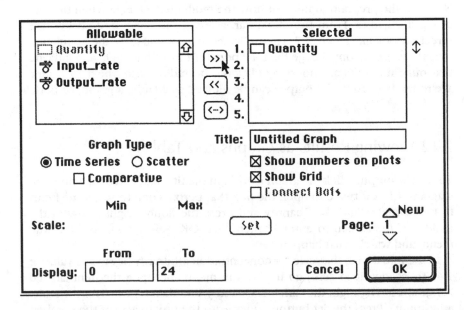

FIGURE 7.8

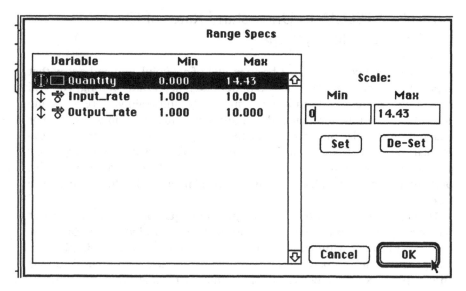

FIGURE 7.9

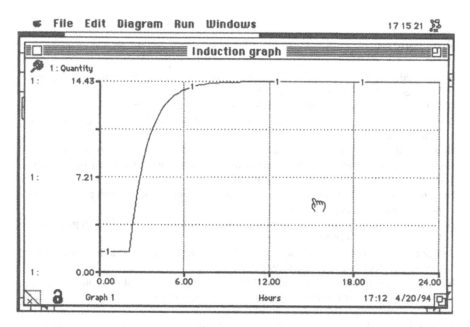

FIGURE 7.10

7.14 Use **STELLA**® to Understand Rates of Change in More Complex Systems

You have sketched in the most basic model of accumulation and decay that is possible. In any applications you undertake in a research setting, you will be testing a system with multiple components, and plainly there are many possible ways by which material or information may flow in a system. For example, molecular biologists study the flow of information from an extracellular signal to DNA to mRNA to protein, and nutrition scientists may evaluate the flow of nutrients from the gut to the blood to the tissues, with possibilities for metabolism, storage, and elimination in the urine or the stool.

The major conclusion from the simple model just used is that the rate at which a system approaches a new steady state is determined primarily by the rate of elimination within a compartment. Is this only true for a simple, one-compartment model, or would it also hold for models with multiple inputs and outputs, exchange between compartments, or multiple compartments? Stay alert to potential answers to this question in subsequent chapters in this text.

References

Berlin, C.M, and R.T. Schimke. "Influence of turnover rates on responses of enzymes to cortisone." *Mol. Pharmacol.* 1 (1965): 149–156.

Gleick, J. *Chaos, Making a New Science.* New York: Penguin Books, 1987.

Godfrey K., ed. In: *Compartmental Models and Their Applications.* London: Academic Press, 1983.

Goodwin, B.C. *Temporal Organization in Cells.* London: Academic Press, 1963.

Green M.H, J.B. Green. "The application of compartmental analysis to research in nutrition." *Annu. Rev. Nutr.* 10 (1990): 41–61.

Hargrove, J.L. "Microcomputer assisted kinetic modeling of mammalian gene expression." *FASEB J.* 7 (1993): 1163–1170.

Hoel, D.G. "A simple two-compartmental model applicable to enzyme regulation." *J. Biol. Chem.* 245 (1970): 5811–5812.

Jacquez, J.A. *Compartmental Analysis in Biology and Medicine.* 2nd ed. Ann Arbor: University of Michigan Press, 1985.

Maynard-Smith, J. "Kinetics of chemical reactions. c. The control of protein synthesis." In *Mathematical Ideas in Biology.* Cambridge, Great Britain: Cambridge University Press, 1968, 107–116.

Monod, J., A.M. Pappenheimer, and G. Cohen-Bazire. "La cinetique de la biosynthese de la β-galactosidase chez *E. coli* considerée comme fonction de la croissance." *Biochim. Biophys. Acta* 9 (1952): 648–660.

Palmiter, R.D. "Ovalbumin mRNA translation." *J. Biol. Chem.* 248 (1973): 2095–2106.

Pelham, H.R.B., and R.J. Jackson. "An efficient mRNA-dependent translation system from reticulocyte lysates." *Eur. J. Biochem.* 67 (1976): 247–256.

Richter, J.D., and L.D. Smith. "Differential capacity for translation and lack of competition between mRNAs that segregate to free and membrane-bound polysomes." *Cell* 27 (1981): 183–191.

Yagil, G. "Enzyme induction." In: Segel, L.A., ed. *Mathematical Models in Molecular and Cellular Biology*. Cambridge: Cambridge University Press, 1980, 68–83.

8

Chronological Time Versus Physiological Time

Second(s)—The second is the duration of 9,650,763.73 periods of the radiation that corresponds to the transition between the two hyperfine levels of the ground state of the Cesium-133 atom.

D.S. Young, "Implementation of SI units for clinical laboratory data. Style specifications and conversion tables. " *Ann. Intern. Med.* 106, 114 (1987).

"I wonder," said Ada, "I wonder if the attempt to discover those things is worth the stained glass. We can know the time, we can know a time. We can never know Time. Our senses are simply not made to perceive it." Vladimir Nabokov, *Ada or Ardor: A Family Chronicle*, McGraw-Hill Book Co., New York, 1969.

If one were to ask a number of people what physiological rhythm they most closely associate with the passage of time, most would probably choose the heart rate. We are consciously aware of this pacemaker in quiet moments just prior to sleep, and we become acutely aware of it during exertion or when frightened or angry. Most students have probably estimated the number of times the heart will beat during a lifetime, and some people morbidly reflect on how many heartbeats have passed and how few remain.

Like other variables, heart rate can be scaled allometrically as a function of body weight. Professor Max Kleiber (1932) derived a relationship using data obtained from several large mammals, and stated that the pulse rate could be estimated as follows:

$$P = 186 \times W^{-1/4} \tag{8.1}$$

where P, pulse rate, is given in beats per minute, and W, body weight, is in kg. In his 1975 book, *The Fire of Life*, Kleiber argued on theoretical grounds that *metabolic time* should be related to weight in the same manner. Before discussing the derivation and important implications of this relationship, let us reiterate that an ordinary person has little concept of the meaning of body weight raised to the −1/4 power. In a moment, we shall use **STELLA®** to deal with this problem and the related problem of metabolic body size.

8.1 Metabolic Pools and Turnover Time

The human body automatically maintains a relatively constant amount of many crucial substances, and should it fail to do so, we would sicken or die. Because all constituents of the body are constantly subject to loss or oxida-

tive destruction, this means that mechanisms must exist to replenish or repair these constituents. Notable examples include oxygen, water, glucose,
adenosine triphosphate, and calcium ion, not to mention blood proteins and
blood cells. For each of these substances, it is possible to measure the amount
present in the body, the amount used up per minute, hour, or day, and the
amount that must be replaced. Yet the quantities present and rates of use differ for every substance, and therefore the amounts required differ.

Biologists customarily name the quantity of each individual substance present in the body as the *metabolic pool* of that substance, and the processes of
utilization and replenishment as the *turnover* of the pool. The total pool
available in the body may be subdivided into *compartments*, each of which
contains some of the material that may be separated by a barrier from the rest
of the whole body pool, or may be chemically distinct. The simplest example
may be the pool of body water, which equals about 60% of the total body
weight, or 42 L for a man with a body mass of 70 kg. From the outside, it is
easy to observe the amount of water taken in as beverages and food (about
2.5 L per day), and the equal amount lost by urination, excretion, perspiration, and evaporation into the air. Yet we know that internally, water is present in the blood plasma, the extracellular fluid, and the intracellular fluid, and
that these are separated from one another by blood vessel walls and by the
plasma membranes of cells. A steady state exists when the amount added to
the pool each day equals the amount lost or destroyed.

Any student who wishes to understand time-dependent processes should
be familiar with the concept of turnover, for it applies to organs, cells, and
cellular components such as enzymes and messenger RNAs, nutrients, and
drugs. The terms in Table 8.1 are used to describe this process.

The terms listed in the table are interrelated in the following ways:

$$k_t = \frac{R_t}{A_t} = \frac{0.693}{t_{1/2}} \tag{8.2}$$

$$t_t = \frac{A_{ss}}{R_t} = \frac{1}{k_t} \tag{8.3}$$

$$t_{1/2} = \frac{0.693}{k_t} \tag{8.4}$$

$$t_t = \frac{t_{1/2}}{0.693} = 1.44 \times t_{1/2} \tag{8.5}$$

TABLE 8.1

Term	Symbol	Definition
Pool size	A_t	Total quantity of a substance present at time t
	A_{ss}	Total quantity present at steady state
Turnover rate	R_t	Total quantity entering or leaving pool per unit time
Fractional turnover rate	k_t	Ratio of R_t to A_t; fraction of pool that is replaced in a specified period (per hour, per day, etc.)
Turnover time	t_t	Time needed to replenish the total pool
Half-time	$t_{1/2}$	Time needed for pool size to increase or decrease by half

8.2 The Metabolic Rate in Mammals Is Related to the Rate of Heat Loss

The processes that support metabolism and turnover produce heat, indeed most of the energy expended as a result of metabolic processes is lost as heat, and is used chiefly to maintain our internal temperature. Humans and other mammals maintain body temperatures very close to 37°C, a value that is usually higher than ambient temperature (typically, 10°-30°C in temperate zones). Because the rate of cooling is proportional to the surface area and the difference between our internal temperature and the ambient temperature, the amount of heat needed to maintain a constant internal temperature is proportional to our surface area. Early attempts to model heat loss simplified the shape of animals and humans, and assumed that surface area was similar to that of a sphere and should be proportional to the 2/3 power of body mass. However, Kleiber (1932) analyzed data for basal metabolic rates of a number of species as reported by other laboratories, and observed that the value was closer to the 3/4 power of the body weight, with a constant scaling factor of about 70. In the studies cited, the average weights of human males and females, respectively, were 64.1 and 56.5 kg, and their basal metabolic rates were 1,632 and 1,349 kcal per day. In comparison, male and female rats weighed 0.226 and 0.173 kg, respectively, and their basal metabolic rates were 25.2 and 20.2 kcal per day.

The regression formula that Kleiber derived to fit these data is:

$$M = 70.3 * W^{0.75}$$

where M is the *basal metabolic rate* in kcal/d and W is body mass in kg. Note that one would need to multiply the predicted values times an activity factor to obtain a 24 h energy expenditure.

8.3 Use **STELLA®** to Solve the Kleiber Equation for Basal Metabolic Rate

The model shown in Figure 8.1 was generated to solve this problem and also to permit comparison of metabolic rate with metabolic time or predicted life span. Instead of using the calculation interval as time, the flow into the Body Mass stock simply increases the mass by weight units. The converters then calculate the 1/4 and 3/4 powers of the mass. The standard metabolic rate (a basal value, not corrected for activity) is then found by multiplying 70 by the value of mass raised to the 3/4 power. The graph shown in Figure 8.2 shows the relationship between basal caloric needs and body mass. Observe that this is not a linear relationship; the *logarithm* of daily caloric needs is proportional to the *logarithm* of the body mass (one log unit represents a power of 10, e.g., $10^2 = 100$).

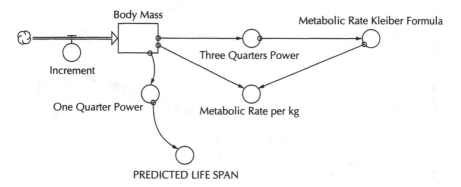

FIGURE 8.1. Diagram depicting a **STELLA®** map that predicts total metabolic rate, metabolic rate per kg body mass, and estimated lifespan based on an allometric equation.

The following equations were entered in the models:

```
Body_Mass(t) = Body_Mass(t - ) dt) + (Increment) * dt
INIT Body_Mass = 0
Increment = IF(TIME < = 1) THEN(0.1) ELSE(0.5)
Metabolic_Rate_Kleiber_Formula = Three_quarters_power
* 70
Metabolic_Rate_per_kg = Metabolic_Rate_Kleiber_
Formula/Body_Mass
One_quarter_power = Body_Mass^0.25
PREDICTED_LIFE_SPAN = One_quarter_power * 11.6
Three_quarters_power = Body_Mass^0.75
```

In comparing metabolic rates of different animal species, the total metabolic rate is not a useful measure. Instead, it is necessary to divide the total metabolic rate by the body mass in order to derive the metabolic rate per kilogram of tissue. This value represents the *specific metabolic rate*, which is expressed in units of energy expended per kg per day under standard conditions. By dividing by the caloric equivalent of oxygen (about 5 kcal per L of oxygen consumed), the specific metabolic rate may be converted to oxygen consumption per kg of tissue per day. The distinction between total metabolic rate and specific metabolic rate becomes clear if one compares the rates estimated for humans of different weights (Fig. 8.3). The total metabolic rate increases with size because it reflects total tissue mass, whereas the specific metabolic rate decreases as mass increases. Adults consume only about half as much oxygen per kg tissue as do infants.

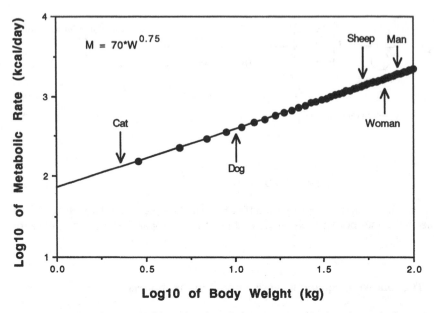

FIGURE 8.2. **STELLA®** was used to solve the Kleiber equation for basal metabolic rate as a function of body mass. The semi-logarithmic plot shows that the estimated rates increase as a power function of body mass, with a constant slope of about 70 kcal per kg.

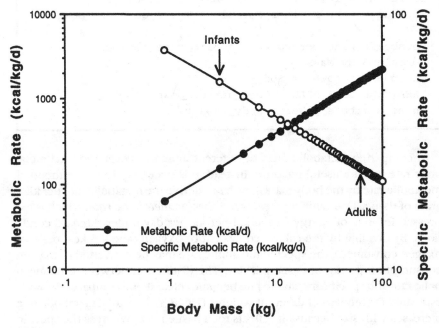

FIGURE 8.3. The distinction between total metabolic rate (kcal per day) and specific metabolic rate (kcal per kg per day) is demonstrated using output from a **STELLA®** model. As body mass increases, the total metabolic rate increases, whereas specific metabolic rate decreases.

8.4 Specific Metabolic Rate and Physiological Time

There is no single definition of time that fits all living creatures. For instance, the passage of one day is not the same for a hummingbird or a mouse as it is for a human or an elephant; by the time the human infant is able to run and talk, the mouse will have matured sexually, produced offspring, and be senescent, if not dead. This example shows that rates at which metabolic events and cellular processes take place in the human body are only indirectly related to clock time or calendar time. The concept of physiological time derives from the need to compare metabolic events among species with very different specific metabolic rates.

The simplest method of relating physiological time to body mass was given by Max Kleiber (1961) in his book, *The Fire of Life*, and restated by Lindstedt and Calder (1981). The relationship was derived on the basis that the total amount of any material in the human body (the pool, P, of the material of interest) must be proportional to body size (or *volume*). Material in the pool equals a scaling fraction multiplied by mass, or kM. Keeping track of the exponents, one would express pool size as:

$$P = k_1 M^1$$

But how can metabolic rate be expressed? Kleiber analyzed data from many different kinds of animals, and observed that metabolic rate was approximately proportional to the 3/4 power of body weight. He argued that the transfer rate (r) of material into and out of the metabolic pool then must also be proportional to the 3/4 power of body mass, that is,

$$r = k_2 M^{3/4}$$

To focus on the most elemental aspect of the relation between transfer rates and body mass, just recall why one must breathe—the answer is to deliver oxygen to the mitochondria in support of cellular respiration. The amount of oxygen that must be transferred to the tissues is directly proportional to the amount of energy being produced in the tissues, so our oxygen needs must also be proportional to $M^{3/4}$. This equivalence between oxygen needs and energy expenditure is called the *caloric equivalent of oxygen*, and it is a number close to 5 kcal per L of oxygen utilized. The exact value ranges from about 5.0 shortly after a mixed meal, to about 4.7 when a person is fasting.

If the turnover time for the metabolic pool is the ratio of the pool size to the transfer rate, then this ratio is one measure of physiological time as it applies to metabolically active organs:

$$t = P/r = k_1 M^1 / k_2 M^{3/4} = k * M^{1/4}$$

As Lindstedt and Calder (1981) pointed out, many physiological functions have a time scale that is close to $M^{1/4}$. We will use **STELLA**® to find out what this means in a moment. However, it is reasonable to ask how much metabolism a tissue is capable of carrying out during a lifespan. Is there a limit, or might we hope to go on forever? The equations of scale suggest that if specific tissue metabolic rate is proportional to $M^{3/4}$ and physiological time is proportional to $M^{1/4}$, then the total amount of metabolism a tissue can carry out in a lifetime can be expressed as follows (Hainsworth, 1981):

$$\text{Lifetime metabolism} = k(M^{3/4})(M^{1/4}) = kM^{1.0}$$

where k is a proportionality constant. This equation can be interpreted to mean that *lifetime metabolism* is directly proportional to mass, or that a tissue is only capable of a certain amount of energy transformation before it ceases to function. If this proves true, then a small mammal with a high metabolic rate per kg weight should have a shorter lifespan than a larger mammal with a lower specific metabolic rate.

A word of caution is merited. The idea that the metabolic rate per kg of tissue is proportional to $M^{3/4}$ is an observation, not an explanation. The observation was originally made in an attempt to relate metabolic rate to heat loss through the surface of the body in many different animals. If this had been true, it was thought that metabolic rate should have been related to mass raised to the 2/3 power (0.67). Though the difference is small, there is no generally accepted reason that explains the discrepancy, which means that this observation is not constructed on a very solid foundation. Even so, if one can accept the idea that specific metabolic rate per kg of tissue is proportional to a number between $M^{2/3}$ and $M^{4/5}$, then physiological time should be proportional to a number between $M^{1/3}$ and $M^{1/5}$.

Finally, suppose that metabolic rate was directly proportional to mass. Then the specific metabolic rate would be proportional to $M^{1.0}$, and the observant student has already noticed that the reciprocal is M^0. This would not mean that time is indefinitely short; it means that metabolic time is independent of mass. But this is preposterous! Every student remembers various guppies and puppies that they raised as children and that have long since expired! There is a nugget of truth in the notion of physiological time, and it is worth using **STELLA®** to try to grasp the truth through the haze of these equations.

8.5 Use **STELLA®** to Compare Scaling for Metabolic Time and Metabolic Rate

Create the model shown (or open the model included with this volume) and run the model with the simulation time (here functioning as a mass increment) to calculate the values corresponding to mass raised to the 0.25 and 0.75 powers. This has been done and plotted as a log-log graph in Figure 8.4. Notice that if this theory holds, the metabolic rate should increase by a factor of 40 as the mass increases from 1 kg to 100 kg, whereas metabolic time should increase by a factor of about 4. If this is true, it could mean that small animals perceive a given time interval as being longer than do large animals, and this could well be true. A rat's heart typically beats 300–400 times per minute, and an adult human's typically beats about 70 times per minute. Human infants have much higher heart rates than their parents do.

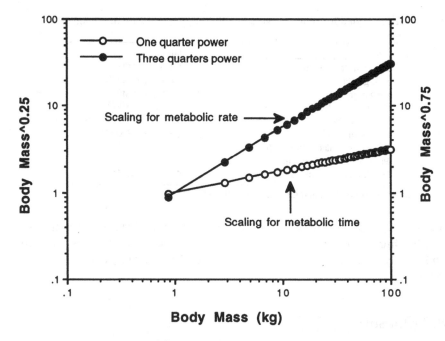

FIGURE 8.4. A **STELLA**® model was used to show the difference between two processes that are both proportional to body mass but that increase at different rates. Metabolic body size increases to the 0.75 power of body mass, whereas metabolic time is hypothesized to increase to the 0.25 power of body mass.

In any event, it seems plausible to assume that one's perception of time could be related to metabolic requirements—a minute can seem like a very long time to a person holding his breath! Lindstedt and Calder (1981) stated this explicitly, saying "the time animals require to perceive and respond to their surroundings seems to scale to $M^{1/4}$." One wonders whether the notorious differences in time perception between men, women, and children might have to do with this concept of metabolic time! In interpreting these curves, it is well to remember that there is also a scaling factor; the proportionality only indicates a general trend.

Some other time-dependent relationships are shown in Table 8.2. In general, these functions can be scaled with an exponent close to 0.25.

8.6 Effects of Body Size on Drug Dosing

Notice that some of the functions that scale to a power of about 0.25 include heart rate, circulation of the blood, and drug metabolism. This scaling has great importance to clinical medicine, because body size is an important aspect of drug metabolism. When drugs are being developed, much of the early testing for safety and efficacy is done in animals, and doses must then be adjusted to humans for clincal trials. Scaling on the basis of body mass is an

TABLE 8.2 Equations relating cycle length (minutes) to body mass (kg) in mammals.

Function	Equation
Life span in captivity	$6.1 \times 10^6 \, M^{0.20}$
50% growth time	$1.85 \times 10^5 \, M^{0.25}$
Gestation period	$9.40 \times 10^4 \, M^{0.25}$
Time to metabolize fat stores	$1.70 \times 10^2 \, M^{0.26}$
Cardiac cycle	$4.15 \times 10^{-3} \, M^{0.25}$
Respiratory cycle	$1.87 \times 10^{-2} \, M^{0.26}$
Time to circulate blood volume	$3.5 \times 10^{-1} \, M^{0.21}$
Plasma clearance, inulin	$6.51 \times M^{0.27}$
Half-life of drug (methotrexate)	$5.8 \times 10^1 \, M^{0.19}$

Source: Lindstedt, S.L., and Calder, W. A., III (1981)

important aspect of this process (Boxenbaum, 1986). The same is true of treating human patients with various medications, because doses given to children by pediatricians are substantially different from those needed by adults.

8.7 Questions

What are the corollaries to the conclusion that metabolic rate and perception of time may be scaled to body mass? If metabolic rate can be reduced to a lower level in our organs by caloric restriction (Masoro and McCarter, 1990), should this promote longevity? Or do other explanations not related to de-creased metabolic rate underlie this observation (McCarter and Palmer, 1992)? Notice, however, that the shape of the curve relating metabolic rate to body mass suggests that rats have more potential life to gain from caloric restriction than do humans (Fig. 8.5). Similarly, do hibernating animals live longer if they reduce core temperature, as we assume if we elect to store our bodies by cryogenic means for space travel? What is the mechanism—is it improved coupling efficiency and enhanced antioxidant status? Do turnover times for specific components vary? How do metabolic rates compare for cells in culture from animals of different specific metabolic rates? Is the relationship lost in culture?

What is the effect of efficient insulation? Do women live longer because their heat production is lower and metabolic rates are more efficiently coupled? Is the decrease in lifespan due to excessive food intake proportional to the excess oxidation?

8.8 A Natural Experiment

If the metabolic rate is scaled according to the Surface Law, then the question of what happens to infants on the day they are born becomes a very interesting one. Try using the **STELLA**® model to estimate the caloric require-

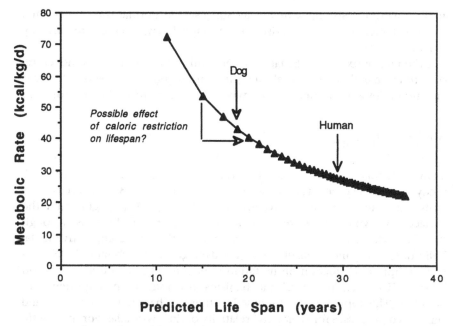

FIGURE 8.5. The prediction that metabolic time is related to body mass raised to the 0.25 power suggests that lifespan should scale similarly. Predictions made on this basis are shown here. On this basis, the lifespan of a dog is estimated appropriately, but the lifespan of a human is underestimated. Might the length of time spent prior to sexual maturation be related to this disparity?

ments (total and per kg weight) of a 60-kg woman before pregnancy and at the end of gestation, when her weight has increased to about 72 kg. Now calculate the metabolic rate of a newborn baby of 4 kg on the basis of the surface law. Should the fetus have the same metabolic rate per kg as the mother before birth, and a metabolic rate appropriate to the Surface Law after being born? Are relevant data available in the medical literature, and can you present arguments in favor of each position?

A related question is whether cells should have the same metabolic rate in the body as they do after isolation and growth in tissue culture, where they are maintained artificially at 37°C by being placed in an incubator with growth medium. If isolated cells maintained the characteristics needed to support energy expenditure in the whole animal, cells from rodents should have a much higher metabolic rate when cultured than do human cells. However, if the cells adapted to the artificial environment, they might all have the same metabolic rates when cultured in artificial medium at constant temperature. Booth et al. (1967) evaluated this concept, using liver cells isolated from humans and rats, and found that rat hepatocytes consumed 2–4 times more oxygen than did human cells in identical medium. Evidence suggested that the cells had the same total capacity for oxygen consumption, but the human cells metabolized nutrients at a lower rate. Data for cells from other animals were consistent, which may suggest that normal (non-cancerous) cells retain metabolic characteristics of the animal in which they are normally found.

If the rate of oxidative damage to DNA and to mitochondria is indeed related to the metabolic rate, it should therefore be possible to show different mutation rates among species even when their cells are studied in isolation.

8.9 Conclusions

Many metabolic functions scale to exponents of body mass. Functions directly related to volume tend to scale to powers of about 3/4, whereas functions related to turnover scale to powers of about 1/4. This originated in the Surface Law, which attempted to relate metabolic rate to heat loss through the skin surface. The idea has important implications for comparative metabolic rates, drug metabolism, oxidative damage to mitochondrial DNA, effects of lipid peroxidation on mutation of genomic DNA, and ultimately on lifespan. However, in general, mechanisms that underlie the exponents are not clear. This topic will be discussed further in Chapter 13 on growth and aging, two processes for which the relativity of time is crucial. For in growth, reproduction, and aging, it is essential to "make hay while the sun shines," an aphorism taken quite seriously by animals and plants living in the Arctic!

Set aside the question of exponents and scaling for different functions. Contemplate the idea that our sense of time may be linked to our metabolic rate, which in turn is partly related to the rate of heat loss through our skin. It may be feasible to use **STELLA®** to frame hypotheses about this overlooked aspect of life, and to learn more about a process that is difficult to understand partly because it is not a linear function of our size, but rather a logarithmic function of mass. Related to the way we experience the passage of time is the way we age, which has practical significance.

References

Booth, J., E. Boyland, and C. Cooling. "The respiration of human liver tissue." *Biochem. Pharmacol.* 16 (1967): 721–724.

Boxenbaum, H. "Time concepts in physics, biology, and pharmacokinetics." *J. Pharmaceut. Sci.* 75 (1986): 1053–1062.

Calder, W.A., III. *Size, Function, and Life History*. Cambridge, Mass.: Harvard University Press, 1984.

Hainsworth, F.R. *Animal Physiology: Adaptations in Function*. Reading, Mass: Addison-Wesley, 1981, 170.

Kleiber, M. *The Fire of Life*. New York: John Wiley, 1961.

Lindstedt, S.L. and W.A. Calder. "Body size, physiological time, and longevity of homeothermic animals." *Quart. Rev. Biol.* 56 (1981): 1–16.

Lynn, W.S., and J.C. Wallwork. "Does food restriction retard aging by reducing metabolic rate?" *J. Nutr.* 122 (1992): 1917–1918.

Masoro, E.J., and R.J.M. McCarter. "Dietary restriction as a probe of mechanisms of senescence," In: *Annual Review of Gerontology and Geriatrics* (Christofalo, V.J., and M.P. Lawton, eds.) Vol. 10. New York: Springer Publishing, 1990, 183–197.

McCarter, R.J., and J. Palmer. "Energy metabolism and aging: a lifelong study in Fischer-344 rats." *Am. J. Physiol.* 263 (1992): E448-E452.

McMahon, T.A. "Scaling physiological time." *Lectures on Mathematics in the Life Sciences* 12: (Amer. Math. Soc. 1980), 131–163.

Pearl, R. *The Rate of Living.* New York: Alfred Knopf, 1928.

9

Energy Needs for Work

"I was working then at the old Sorbonne, in an ancient laboratory that
opened into a gallery full of stuffed monkeys."
Jacques Monod, *Nobel Lectures in Molecular Biology* (1965)

Even though the work of performing genetic experiments differs from heavy
physical labor, there is still an unstinting need to spend intensive hours in the
laboratory to be productive. Certainly, Monod would have understood that
there are four kinds of work for which food energy is used: mechanical,
chemical, electrical, and osmotic. If the thesis is true that there is a connection between work and our sense of time, it is useful to ask how **STELLA**®
might help one to understand the concepts of energy balance. Energy balance occurs when the amount of food energy we take in equals the amount
of work our bodies perform, irrespective of the nature of our labors. Therefore, let us contemplate half of this equation.

Most of the heat production that causes us to have stable internal temperatures, and to give off heat through our skins and in the warm air we exhale
with each breath, results from the inefficiencies inherent in doing work. The
energy used to accomplish work is entirely taken in as chemical energy in
foods, in the form of covalent bonds between atoms in the fats, carbohydrates, proteins, and alcohol that we ingest. Americans are accustomed to
thinking of food energy in terms of the *calorie* (the kilocalorie of science),
which is defined as a unit of heat. However, the committee that defined the
standard units of measure recommended the *joule* as the common unit of energy (1 cal = 4.184 J), on the basis that all forms of energy are equivalent. The
metabolic machinery of the body must use the chemical energy of foods to
accomplish the work of manual labor, the work of the heart and neurotrans-
mission with exchange of sodium and potassium ions across membranes, and
the work of circulating the blood and concentrating the urine. It is all these
tasks that cause our bodies to become warm, so let us warm to the task of using **STELLA**® to come to grips with biological work. The program will assist
us in comprehending the mysteries of interconversion of energy.

9.1 Human Work Follows the Laws of Thermodynamics

Pick up the Sunday newspaper supplement on nearly any weekend, and you
will find an advertisement stating that another European scientist has discovered a magic, fat-burning formula, which can be yours for a mere $19.95.

Resist the temptation to buy it! For fat is a stored form of energy, and there are only two means to remove it from the body: (1) oxidation of fatty acids in mitochondria and peroxisomes, and (2) liposuction. So what are these laws of thermodynamics, which interfere with long-standing (if questionable) business practices?

The *first law of thermodynamics* states that when the chemical energy content of a system changes (such as in a human being after a meal), the magnitude of the change must equal the sum of all forms of energy given off or absorbed. This is also called the Law of Conservation of Energy, which is the law that causes food energy absorbed during a holiday feast to be converted to fat during the postprandial football game, and not to move from the hips until the celebrant finally goes to the gym!

The *second law of thermodynamics* is a bit more obscure. Lord Kelvin stated it roughly as follows: It is impossible to take heat from a reservoir and convert it to work in a cyclic process without, in the same operation, transferring heat from a hot to a cold reservoir. Warmth cannot spontaneously move from a cold body to a hot body; this has to do with order and entropy, the archrival of every biological system. In a subtle way, this is the reason that food energy is converted to the common form of high-energy phosphate bonds in adenosine triphosphate (ATP) just prior to use. ATP provides a source of free energy that can be used to create order in the form of new proteins and DNA, but at the same time heat is given off. We give off heat because we must to counter disorder. Lord Kelvin also is reputed to have said, "In the long run, we are all dead," and he would have observed that a cold, dead body can no longer use free energy to counteract entropy! So let's get to the point. However, don't forget that there is no magic fat-removing cream. Better to invest the $19.95 at compound interest and either pay your gym fees or save up for the liposuction!

STELLA® can be used to become familiar with work whether expressed as kcal or kj, and rate of work expressed as watts (j/s). Fig. 9.1. depicts a control panel that operates the model for calculating work done in stair-

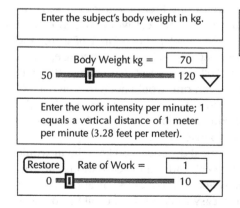

Enter the subject's body weight in kg.		Then run the program using the Run menu. Set the Time Spec to the number of meters you wish to travel.
Body Weight kg = 70		
50 ▭ 120 ▽		Cumulative Verti. . . 240.0
Enter the work intensity per minute; 1 equals a vertical distance of 1 meter per minute (3.28 feet per meter).		Work as heat ki. . . 39.3
		Mechanical work. . . 164.5
Restore Rate of Work = 1		Work per minute. . . 3.9
0 ▭ 10 ▽		Watts 274.1

FIGURE 9.1

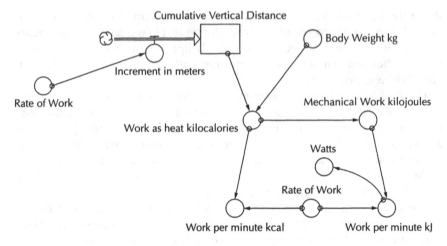

FIGURE 9.2

climbing or ascending a specified vertical distance as a stipulated rate, based on the model of Fig. 9.2. To run the model, enter a subject's body weight in kg by using the slider shown in Fig. 9.1. Then enter the rate of work in units of vertical meters per minute. In this case, increments of total vertical distance are entered using simulation time from the Time Specs feature of the Run menu. For instance, setting Time Specs to a value of 100 is like asking a subject to ascend a vertical distance of 100 meters, or over 330 feet. This is not quite as high as the Washington Monument! Interested students might wish to calculate the minimum work needed to climb various mountains wearing backpacks of different weights.

9.2 Human Mechanical Work

The familiar (kilo)calorie defines a measure of heat, specifically the amount of heat required to raise the temperature of one liter of water by 1°C. However, there is an equivalence of heat and mechanical work, which can be shown by placing a thermostat in a closed, insulated container of water and shaking vigorously. This causes the temperature of water to increase, and it is not due to cold fusion—it is due to "elbow grease!" To change the temperature of a liter of water by 1°C, however, one would need to attach a 427 kg weight to a pulley system and thence to a stirrer in the water, and drop the weight 1 meter. Yes, the mechanical equivalent of 1 kcal is 427 m * kg—no wonder it is easier to gain weight than to lose it!

To make it simpler, ask how much vertical distance a 60-kg woman would need to climb on stairs to expend one kcal under ideal conditions (no friction, perspiration, etc.). The answer is about 7.1 meters, or about twenty-three *vertical* feet. Given that it is very easy to consume a meal of 500 kcal, one begins to see why obesity has become a national problem! Under ideal conditions, our subject would expend 500 kilocalories after climbing perhaps 10,000 vertical feet!

Let us set up a simple **STELLA®** model to calculate mechanical work in terms of kilocalories and convert that to the official unit, the joule (1 kcal = 4.184 joules). I must confess that I always rooted for the erg as the official unit, for that is what most individuals say when asked to move the refrigerator! Incidentally Max Kleiber, a master of *calorimetry*, pointed out that Professor Joule became famous for showing that 427 m * kg of work = 1 kilocalorie of heat, and that this highly meaningful relationship must now be replaced by the equation, 4184 J = 4184 J, "a meaningless tautology" (Kleiber, 1975, p. 130). *Calor* means heat in Latin, and the root still connotes something to most of us, unlike the term joule.

Equations for the Mechanical Work Model

```
Cumulative_Vertical_Distance(t) =
Cumulative_Vertical_Distance(t - dt) +
(Increment_in_meters) * dt
INIT Cumulative_Vertical_Distance = 0
Increment_in_meters = Rate_of_Work * 1
Body_Weight_kg = 70
Mechanical_work_kilojoules = Work_as_heat_kilocalories
* 4.18
Rate_of_Work = 1
Watts = Work_per_minute_kJ * 1000/60
Work_as_heat_kilocalories =
Body_Weight_kg*Cumulative_Vertical_Distance/427
Work_per_minute_kcal =
Work_as_heat_kilocalories/Rate_of_Work
Work_per_minute_kJ =
Mechanical_work_kilojoules/Rate_of_Work
```

9.3 The Osmotic Work of the Kidney

The transport of substances against a concentration gradient requires energy in the form of osmotic work. The kidney is one of the most metabolically active organs, and one that is quite sensitive to damage because of the work it does to recover all nutritionally important substances from the kidney tubules after the low molecular weight material passes through the glomerular filtration barrier. It must actively reclaim glucose, amino acids, vitamins, and electrolytes from the filtrate, and it also indirectly performs work to excrete nitrogenous wastes such as urea, ammonia, and creatinine. It is an unfortunate aspect of kidney disease that loss of function tends to be progressive, and treatment includes formulation of diets that will minimize the work required, thereby reducing the burden on individual functional units or nephrons.

It is possible to estimate the amount of work done by the kidney; let us create a **STELLA**® model to do this for the most abundant nitrogenous waste, urea. The amount of energy needed to perform this work is called *free energy*, which is the minimum requirement if the body were acting with complete efficiency. Kleiber (1975, p. 391) gives an example of a means to calculate the work done by the kidney. The formula for calculating work required to create a concentration gradient is

$$-\Delta F = nRT * \ln (C_1/C_2) \tag{9.1}$$

where ΔF = change in free energy expressed in calories (not kilocalories!)

n = number of moles of substance transported per time interval
R = gas constant = 1.987 cal/°C
T = absolute temperature in °K = 273 + °C
C_1 = lower concentration (in blood) from which substance is transported
C_2 = higher concentration (in urine) to which substance is transported.

The example given by Kleiber (1975) involved an experiment in which urea was fed to men and concentrations in the blood and urine were measured; under these conditions, the investigators estimated that urea transport accounted for 77% of the work needed to form the urine. As an aside, it should be noted that these early experiments were done before the involvement of the osmotic gradient in the renal medulla and the process of countercurrent concentration were understood. It might be better for us to use **STELLA**® to calculate the work done in forming the medullary gradient, but let us nevertheless use Kleiber's example.

Urea is generated from the free amino acid pool in the body as a result of amino acid catabolism. One of the key elements in treating a patient with progressive kidney disease is to minimize the production of urea by using low protein diets. The special diets contain as little as 0.28 g protein/kg body weight/d as compared to the RDA value of 0.8 g/kg/d and the usual intake of about 1.0–1.2 g/kg/d in the United States (Kopple, 1988). Therefore, the model shown in Figure 9.3 was set up in a manner that reflects the work done by the kidney in excreting urea formed from dietary protein.

The minimum protein requirement as measured in nitrogen balance studies is about 40 g protein per day, which corresponds to 40 g protein/6.25 g N (nitrogen) per g protein = 6.4 g urea N per day. Urea, $CH_2(NH_2)_2$, has a molecular weight of 50 and is 56% N by weight. Therefore, excreting 6.4 g of urea N is equal to 11.4 g of urea per day. This gives us an approximate way of predicting how much urea will appear at higher protein intakes; the conversion factor is about 6.4 g urea N/40 g protein = 0.16 g urea N per g pro-

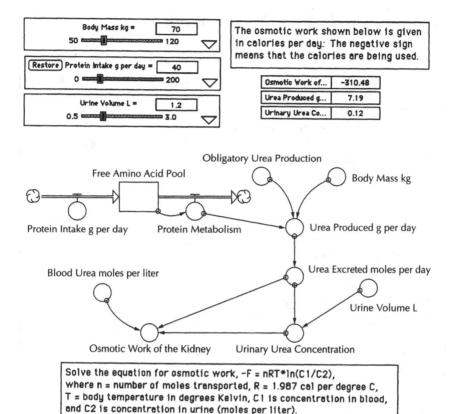

FIGURE 9.3

tein, or 0.285 g of urea per g protein. Under most conditions, adults stay in protein balance, showing a loss of a few percent of lean body mass per decade of life after 30; this can be ignored for a simulation of a day's balance.

The model shown in Figure 9.3 was created to calculate urea formation and work needed to excrete urea for various protein intakes, assuming an initial free amino acid pool of 100 grams, and an arbitrary half-life of 0.5 days for the free amino acid pool. This value equals a fractional exchange rate of $0.693/0.5 = 1.386$ d^{-1}. Urine volumes, initial body weight, and protein intake can be specified, after which the model calculates the amount of urea excreted and the amount of work done by the kidney. The value is not large, but remember, this result assumes perfect efficiency. The original experiment mentioned by Kleiber (1975) suggested that the kidney functioned at about 2% efficiency. If this is true, the calculated value should be multiplied by 50 to calculate the energy requirement for urea transport, which is a significant component of the daily energy budget.

Equations Used for the Osmotic Work Model

```
Free_Amino_Acid_Pool(t) = Free_Amino_Acid_Pool(t - dt)
+ (Protein_Intake_g_per_day—Protein_Metabolism) * dt
INIT Free_Amino_Acid_Pool = 100
Protein_Intake_g_per_day = 40
Protein_Metabolism = Free_Amino_Acid_Pool * 1.386
Blood_Urea_moles_per_liter = 0.0036
Body_Mass_kg = 70
Obligatory_Urea_Production = 0.037
Osmotic_Work_of_the_Kidney =
Urea_Excreted_moles_per_day*1.987 * 310 *
(LOGN(Blood_Urea_moles_per_liter)-
LOGN(Urinary_Urea_Concentration))
Urea_Excreted_moles_per_day =
Urea_Produced_g_per_day/50
Urea_Produced_g_per_day = Body_Mass_kg *
Obligatory_Urea_Production + 0.115 * Protein_Metabolism
Urinary_Urea_Concentration =
Urea_Excreted_moles_per_day/Urine_Volume_L
Urine_Volume_L = 1.2
```

The problem in kidney disease originates from the fact that no new nephrons are formed after we are born. And although the kidney has enormous reserve capacity, as each nephron is lost, the workload on the remaining nephrons increases. (An excellent use of **STELLA**® would be to model this vicious cycle.) Extreme workloads are produced in diseases such as diabetes mellitus, in which glucose concentration exceeds the capacity for reabsorption, and the sugar is lost into the urine. In addition to the higher energy expenditure supporting glucose recovery, the high plasma glucose levels produce a chemical reaction with constituents of the filtration barrier, and result in thickening of the membranes in Bowman's capsule. The progressive nature of kidney disease is an example of a vicious cycle, which is also amenable to modeling with **STELLA**®! Students may wish to review the medical literature to determine whether useful models of kidney deterioration have been described, and if not, to remedy the situation!

9.4 Conclusions

STELLA® can be used to calculate the use of food energy to do the work required to maintain the human body or to perform athletic feats or different kinds of manual labor. The software can estimate energy that is lost as heat under different circumstances, and can illustrate other concepts that underlie the field of bioenergetics. Important topics that students may wish to explore in the medical literature include the effect of diabetes mellitus on kidney function, the effect of dietary protein on kidney function, and aspects of wasting that occurs in fever, after surgery, and in diseases such as cancer, alcoholism, and AIDS-related complex.

References

Kleiber, M. The Fire of Life. An Introduction to Animal Energetics. Huntington, N.Y.: Robert E. Krieger Publishing Co. (1975).

Kopple, J.D. (1988) "Nutrition, Diet, and the Kidney." In: Shils, M.E., and V.R. Young, Eds. *Modern Nutrition in Health and Disease*, 7th ed., Philadelphia: Lea and Febiger, 1230-1268.

Munro, H.N., and M.C. Crim. (1988) "The Proteins and Amino Acids." In: Shils, M.E., and V.R. Young, Eds. *Modern Nutrition in Health and Disease*, 7th ed., Philadelphia: Lea and Febiger, 1-37.

Nobel Lectures in Molecular Biology, 1933-1975, New York: Elsevier North-Holland, 1965.

10

The Human Thermostat

"It has been suggested that two types of human beings may be distinguished by pattern of their temperature fluctuations during a day: the early risers and the late risers. The early risers have a relatively high body temperature in the morning and are barbarically cheerful before breakfast."

Max Kleiber, *The Fire of Life*, 1975, p. 152.

The modern denizen of a suburban home with central heating and air conditioning probably has little idea of what thermodynamic feats the human body can demonstrate. While travelling in Tierra del Fuego on a bitterly cold winter's day, Charles Darwin was astonished to see a woman, completely without clothing, nursing an infant and apparently untroubled by the cold sleet that was falling (Dibner, 1964). Geographers understand that the fires referred to in the name, Tierra del Fuego, are volcanic, and have nothing to do with the air temperature of that grand but inhospitable locale! Even today, there are societies in which people sleep exposed to the sky with little or no clothing, warmed if at all by small fires. The Australian aborigine and the Kalahari bushman are but two examples. Suffice it to say that intrepid scientists have demonstrated, by means of rectal thermometers, that the human body cools during such exposure, but extremes that would be lethal to a city dweller are tolerated as a matter of course among those who live under less temperate conditions.

10.1 Heat Production and Body Temperature

The basal metabolic rate suffices to maintain our body temperature near 37°C during the inactive period of sleep, although there is generally a small decline over the course of the night. As the surrounding air temperature changes during the seasons, most people adjust their comfort zone by adding or removing blankets, so that we are able to maintain body temperature over a broad *zone of thermoneutrality* that extends from about 20°C to 35°C. The lower limit at which body temperature can be maintained without increasing its heat production rate above the basal level is called the critical temperature. Below this level, temperature is maintained by increased heat production from one of three physiological mechanisms: endocrine, neural, and shivering thermogenesis. Nonshivering thermogenesis is due to combined effects of norepinephrine on heat production in brown fat tissue, and effects of thyroid hormone on the metabolic rate. Heat production due to shivering is caused by a neural reflex. Most readers can surely recall a time they experienced shivering; two examples that come to my mind were during a fishing trip along the Kenai River in Alaska, and an early spring attempt to qualify for a swimming merit badge in Great Falls, Montana. The pop tune, "Here Comes the Sun," always made me think that one of the Beatles must have been a Boy Scout.

10.2 Modeling the Human Thermostat

The classical law governing heat loss is generally referred to as Newton's Law of Cooling, which states that the rate at which temperature decreases in a warm body in a cold environment is proportional to the temperature difference between the body and its surroundings. Kleiber (1975) icily observed that "Newton's law of cooling applies only to bodies without internal heating, thus only applies to dead bodies," and listed it in his glossary as "Newton's cooling law for carcasses." He also took exception to the alternative, Fourier's Law of Heat Conduction, partly because we lose heat by conduction, convection, evaporation, and radiation. For our purposes, it should be adequate to state that a linear approximation of heat flow can be formulated by lumping all the processes of heat loss (an assumption that is not always justifiable):

$$q = A * (T_b - T_a)/r \qquad 10.1)$$

where q = rate of heat flow per unit time
  A = cross section of area of flow
  T_b = body temperature
  T_a = ambient temperature
  r = specific insulation = L/λ
  L = thickness of insulation
  λ = heat conductivity of insulating material

Note that this equation only states that the rate of heat flow (and resulting change in temperature) is proportional to the difference between the temperature of a body and the temperature of the air, where the proportionality constant equals A/r. This affords an excellent opportunity for a classroom demonstration, since the only items required are a thermostat, a milk bottle or other vessel, and a source of hot water.

To model human temperature regulation, the important point is that an estimate of heat loss be made, and a mechanism be devised to maintain body temperature by increasing the rate of heat production below the critical temperature. In addition, there must be an increase in heat loss in the range between the critical temperature and body temperature, and then an increase in work as a component of heat loss by evaporative cooling.

Modeling Heat Loss with **STELLA®**

To solve the equation for heat loss in humans, first one needs to calculate the surface area through which the heat is lost. A simple device was created with **STELLA®** to do this, and to allow comparison with a cylinder of similar dimensions (Fig. 10.1). Output of the surface area calculator was used to estimate heat loss from a human, using a model set up so that heat loss follows Eq. 10.1. An input for the basal metabolic rate was employed, and heat content of the body was calculated using a specific heat for the tissues equal to 0.83 kcal per kg per degree C (the specific heat of water is 1.0). The model is shown in Figure 10.2. A stock representing environmental temperature was added, and the calculation interval was used to change the temperature in degrees Celsius.

Figure 10.3 shows the estimated heat loss and the rate of heat production above the basal metabolic rate that is predicted by the model. A zone of thermoneutrality is observed between about 35°C and 20°C; the temperature at which heat production must increase to maintain body temperature is called the lower critical temperature (T_c). Below the critical temperature, heat production increases to maintain body temperature until regulation fails (usually due to depletion of energy stores), and we become hypothermic. From about 20°–35°C, the body produces excess heat that is not needed to maintain the core temperature. Changes in blood flow and behavior (such as adjusting the house thermostat) equilibrate heat production and loss, and above that

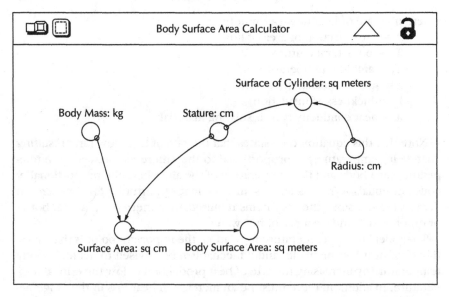

FIGURE 10.1. **STELLA®** device for calculating surface area of the human body and a cylinder of similar dimensions. This is part of the model for estimating the rate of heat loss.

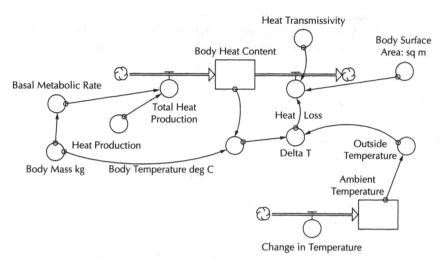

FIGURE 10.2. **STELLA®** Model for calculating the rate of heat loss from the human body. Components include heat loss as determined by the rate of cooling, a device to change the ambient temperature, a converter to calculate body temperature from body heat content, and the basal metabolic rate. Not shown is the set of controls in the high level map that allows the user to change these variables.

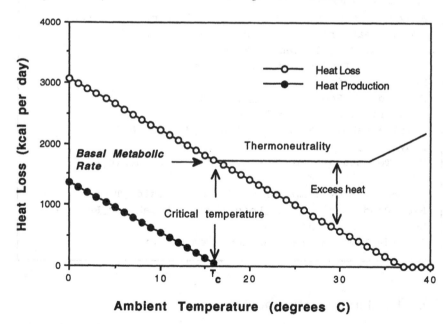

FIGURE 10.3. Predicted rate of heat loss as a function of environmental temperature. The lower critical temperature (T_c) is the point at which basal metabolic rate equals the rate of heat loss; above that temperature, excess heat is generated and behavioral mechanisms are used to compensate. Note that the rate of heat production required to maintain temperature equals zero at the T_c. At a temperature near 30°C, perspiration increases as a means of augmenting heat loss.

range, we begin to perspire and pant to dissipate heat by evaporative cooling. The latter works well on the arid western plains, but not in humid southern climates during late summer, because evaporative cooling depends on the relative humidity.

Equations Used in the Heat Loss Model

```
Body_Mass:_kg = 70
Body_Surface_Area:_sq_m = Surface_Area:_sq_cm/10^4
Radius:_cm = 15
Stature:_cm = 165
Surface_Area:_sq_cm = 71.84 * (Body_Mass:_kg^0.425) *
(Stature:_cm^0.725)
Surface_of_Cylinder:_sq_meters = PI * 2 * (Stature:_cm
* Radius:_cm + Radius:_cm^2)/10^4
Ambient_Temperature(t) = Ambient_Temperature(t - dt) +
(Change_in_Temperature) * dt
INIT Ambient_Temperature = 0
Change_in_Temperature = 1
Body_Heat_Content(t) = Body_Heat_Content(t - dt) +
(Total_Heat_Production - Heat_Loss) * dt
INIT Body_Heat_Content = 2150
Total_Heat_Production = Basal_Metabolic_Rate +
Heat_Production
Heat_Loss = Body_Surface_Area:_sq_m *
Heat_Transmissivity * Delta_T
Basal_Metabolic_Rate = 70 * Body_Mass_kg^0.75
Body_Mass_kg = 70
Body_Temperature_deg_C = Body_Heat_Content/(0.83 *
Body_Mass_kg)
Delta_T = Body_Temperature_deg_C - Outside_Temperature
Heat_Production = Heat_Loss - Basal_Metabolic_Rate
Heat_Transmissivity = 43
Outside_Temperature = Ambient_Temperature
```

10.4 The Human Thermostat

Humans possess a sophisticated thermostat that includes temperature sensors in our skin (Jokl et al., 1992). Our brains sense the rate of heat gain or loss and adjust metabolic heat production and blood flow to the skin to maintain body temperature over a wide range of external temperatures. It is also common experience that we actively make changes if we feel too hot or cold;

men and women differ in this respect (Graham, 1988; Anderson et al., 1995). It is probably not necessary to cite the medical literature to prove this point! Nevertheless, women maintain the same body temperature as men despite a lower rate of heat production, and this may be partly due to the gender-related difference in subcutaneous fat. However, the physiological response of vasodilation and vasoconstriction is probably a more important determinant of our responses to thermal stress (Anderson et al., 1995).

The simple model shown in this chapter allows heat loss to be calculated, but does not attempt to model feedback regulation of human heat production. The interested student is referred to the publications listed in the references for more comprehensive models (Tikuisis et al., 1988; Jokl et al., 1992; Gardner and Martin, 1994; Tikuisis, 1995). Though other references are not listed, there are also many models of temperature regulation in marine mammals, birds, and lizards.

References

Anderson, G.S., R. Ward, and I.B. Mekjavic. "Gender differences in physiological re-actions to thermal stress." *Eur. J. Appl. Physiol.* 71 (1995): 95-101.

Dibner, B. *Darwin of the Beagle*. New York: Blaisdell Pub. Co., 1964, 45.

Gardner, G.G., and C.J. Martin. "The mathematical modelling of thermal responses of normal subjects and burned patients." *Physiol. Meas.* 15 (1994): 381-400.

Graham, T.E. "Thermal, metabolic, and cardiovascular changes in men and women during cold stress." *Med. Sce. Sports Exerc.* 20 (1988): S185-S192.

Jokl, M.V., P. Moos, and J. Stverak. "The human thermoregulation range within the neutral zone." *Physiol. Res.* 41 (1992): 227-235.

Kleiber, M. *The Fire of Life: An Introduction to Animal Energetics*. Huntington, N.Y.: Robert E. Krieger Publishing Co., 1975.

Tikuisis, P., R.R. Gonzalez, and K.B. Pandolf. "Thermoregulatory model for immersion of humans in cold water." *J. Appl. Physiol.* 64 (1988): 719-727.

Tikuisis, P. "Predicting survival time for cold exposure." *Int. J. Biometerol.* 39 (1995): 94-102.

11

Dietary Polyunsaturated Fats and Your Cell Membranes

"I've always suspected that some of the cells in there are fluffing off much of the time, and I'd like to see a little more attention to real work."
Lewis Thomas, M.D., 1974, p. 67

It is often said that "one is what one eats." Think of this in relation to the three macronutrient categories of carbohydrate, protein, and fat. If one eats more carbohydrate, does this change one's body composition in some fundamental way? More glycogen will be stored in muscle and other tissues, but there will be no other fundamental change in body composition. Similarly, if one eats more protein, only a small part of the excess amino acids is stored as labile protein in liver; the rest is quickly used for energy. Again, there is no fundamental change in one's body composition, for the amino acids used to make more proteins are assembled using the genetically encoded templates of messenger RNA, and one's diet does not change that code.

In the case of dietary fats, the situation is very different. It has been demonstrated clearly that the composition of the fats in the body, both in adipose tissue fat stores and in membrane lipids, changes markedly in response to diet. Although this compositional change makes little physiological difference in the fat depots, it significantly affects the availability of lipid precursors that are used for hormonal signals. The aim of this chapter is to explain the concept of the saturation kinetic model for membrane phospholipids, and to provide a tool that allows one to explore the outcomes of changing the diet.

Some of the most important health effects of fatty acids have to do with a form of fat called phospholipid that is an integral part of every cell membrane. The storage form of fat is called triglyceride or triacylglycerol, and is comprised of three fatty acids bonded (esterified) to a three-carbon backbone called glycerol. Triacylglycerol is a neutral, uncharged form of fat. Phospholipid is similar in that it is made of glycerol with two fatty acids esterified in positions 1 and 2, but the third position contains a charged moiety bonded to the backbone through a phosphoester group.

11.1 The Impact of the Food Supply on Proinflammatory Membrane Phospholipids

The studies of Dyerberg and colleagues (1978) on diet and the incidence of cardiovascular disease in Greenland Eskimos suggested that dietary n-3 fatty acids reduced blood coagulation and mitigated coronary artery disease. A broader interpretation, however, is that the balance of polyunsaturated fatty acids (PUFA) in the food supply markedly affects the composition of membrane phospholipids in every organ and cell in the human and animal body. A striking example of this phenomenon is that the food supply of the Inuit is highly enriched in n-3 PUFA. Innis (1989) suggested that this explains the two-fold enrichment of highly unsaturated n-3 fatty acids (HUFA) in red cell phospholipids and the approximately 50% reduction in n-6 HUFA when compared to Caucasians who consume "Western" diets devoid of narwhal, Arctic char, and ringed seal blubber. Moreover, the ratio of arachidonic acid (C20:4 n-6) to eicosapentaenoic acid (C20:5 n-3) is approximately 1:1 in the Inuit, but about 16:1 for age-matched Caucasian males.

Membrane phospholipases are never totally inactive, and there is a continuous release of eicosanoids that increases in response to innumerable stimuli, from toxic or traumatic insult to bacterial and viral pathogens and allergic reactions such as rhinitis and asthma. Also, it is thought that the U.S. food supply now contains much higher levels of linoleic acid, the precursor to arachidonate, and lower levels of α-linolenic acid, the precursor to EPA, than at any previous time in human history (Budowski, 1989). The proportion and types of fat in the American diet have changed due to the destruction of linolenic acid during hydrogenation of oils, the use of corn as a primary feed for livestock, and the availability of high-linoleic acid oils. Increasing the ratio of PUFA to SAFA is recommended as a means of reducing consumption of saturated fatty acids and cholesterol in order to reduce the risk of coronary artery disease. One must ask whether what is currently "normal" is desirable, as Lands and colleagues (1991) put the question. Should Americans who do not have a family history of coronary artery disease follow the same guidelines as those who do?

11.2 Competitive Model for the Effect of Diet on Membrane Phospholipids

It is understood that dietary fat affects cellular membrane phospholipids, and that esterified arachidonic acid (20:4n-6) is the major precursor to inflammatory eicosanoids, which affect the course of many human diseases. Figure 11.1 depicts the pathway by which dietary fats are taken up from lipoprotein particles in the blood as nonesterified fatty acids (NEFA). The shorter-chain linoleic and α-linolenic acids are converted to a fatty acyl coenzyme A

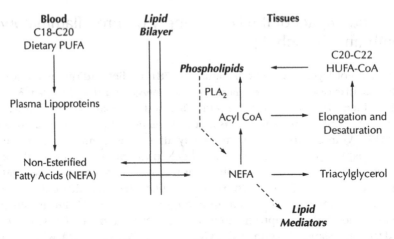

FIGURE 11.1. The basis for the saturation kinetic model of membrane composition is shown here. Dietary lipids are carried to the tissues as components of lipoprotein particles. Lipoprotein lipase releases fatty acids from the particles, and the fatty acids cross cell membranes in the form of nonesterified fatty acids (NEFA). These are either stored in the form of triacylglycerol, or esterified to coenzyme A to form acyl coA. After elongation and desaturation, highly unsaturated fatty acids (HUFA) are formed, including arachidonic acid and eicosapentaenoic acid. These fatty acids are used to produce membrane phospholipids. Inflammatory mediators activate phospholipase A2 and release these fatty acids from the membranes; the free fatty acids are used to generate reactive lipid mediators such as prostaglandins, thromboxanes, and leukotrienes.

(a carrier molecule), and then are elongated and desaturated to produce arachidonic acid, eicosapentaenoic acid, and docosahexaenoic acid. Because membrane lipids are constantly being released, utilized, and replaced, the cell membranes gradually change to reflect current dietary composition. Many anti-inflammatory compounds (such as aspirin, acetaminophen, and glucocorticoids) inhibit production of the eicosanoids, and thereby relieve pain and inflammation. However, it is difficult to make predictions concerning the possible effect of dietary fat on disease processes without reference to a quantitative model. Fortunately, Lands and colleagues (1990, 1991) have provided a quantitative model of the response of membrane phospholipids to dietary fat composition. The model is based on insertion of dietary phospholipids into membranes with *half-maximal responses when the individual polyunsaturated fatty acid (PUFA) is present at about 0.1% of total dietary energy*. Different dietary PUFA compete with one another for a limited number of sites in the membranes with inhibitory concentrations (K_i's) that range from about 0.05 en% to 2.0 en% of total calories.

What do these numbers imply concerning the effect of diet on membrane precursors to the inflammatory eicosanoids? The two important implications are that membranes are *saturated* with individual phospholipids when specific precursors are present at about 1% of total calories. However, the *relative* proportion of specific highly unsaturated fatty acids (HUFA) depends on the relative concentration of the precursor in the diet. Therefore, it is not feasible to decrease the abundance of arachidonate very significantly by changing the amount of oil consumed, even for low-fat rodent diets. However, changing the relative amount of competing n-3 and n-9 PUFA precursors is highly effective in altering the proportions of n-3, n-6, and n-9 HUFA in membranes.

$$n=6 \text{ as } \% \text{ HUFA} = \cfrac{100}{1 + \cfrac{C_6}{en\%6}\left\{1 + \cfrac{en\%3}{C_3} + \cfrac{en\%0}{C_0} + \cfrac{en\%6}{K_s}\right\}} \qquad (11.1)$$

Each of the terms labeled C_n is a constant that was derived to provide best fit of this model to experimental studies. For lipids in liver, the constants are: C6, 0.05 en%; C3, 0.10; C0, 2.0; Ks, 0.3. En%6 and en%3 are the amounts of n-6 and n-3 fatty acids in the diet expressed as a percentage of calories. En%0 is the amount of all other fatty acids (SAFA plus MUFA).

For most readers, this equation and its kin fall in the category, "unknowable to mortals." However, **STELLA®** can be used to solve the equation, and to show that the saturation kinetic model of membrane composition has profound implications for human health. It is also a superb example of the power of quantitative thinking as applied to expermental design, for it allows one to predict the effects of diet on membrane composition. Before we solve this, let us examine some information that explores potentially beneficial effects of *saturated* fatty acids. Heretical? Read on!

11.3 Why Are Beef Fat and Pork Fat Protective Against Alcohol-Induced Liver Disease?

Nanji and French (1985) noted that cirrhosis seldom develops in alcoholics who eat moderate amounts of pork. In pursuing this lead in an animal model, they observed that dietary linoleic acid increased tissue injury during alcohol-induced liver disease compared to diets with equal amounts of tallow or shorter-chain SAFA (Nanji and French, 1989; Nanji et al., 1989). Their protocol involved feeding liquid diets by gastric intubation, with the experimental diets containing 25 en% fat and 36 en% ethanol (replacing an equal amount of carbohydrate). Their studies consistently demonstrate that *ethanol produces little or no liver injury when the diets provide saturated fatty acids, but that massive fatty infiltration, necrosis, and apoptosis occur if the diets contain high amounts of either n6 or n3 unsaturated fatty acids.* The mechanism of injury has been proposed to be elevated free-radical production as a result of induced microsomal cytochrome P450IIE1, and elevated production of inflammatory mediators from membrane-bound phospholipids. In the most severe cases, not only does fatty infiltration occur, but the liver cells also die and become necrotic. Affected cells show abundant smooth endoplasmic reticulum that becomes very dilated. Under these specific circumstances, SAFA are protective and PUFA contribute to necrosis!

11.4 Contradictory Guidelines for Dietary Fat

Because many dietary guidelines attempt to be disease-specific, some contradictions arise. For example, the National Cholesterol Education Panel (NCEP) recommended a Step 2 diet for cholesterol reduction that contains 5 en% saturated fatty acids (SAFA), 10 en% MUFA, and 10 en% PUFA (NCEP, 1993). This represents a change not only from typical earlier diets, which contained much more short and long-chain saturated fatty acids from butter, lard, and tallow, but also from the current American diet, which typically contains about 14 en% SAFA, 14 en% MUFA, and 6 en% PUFA. Current recommendations also include an increase in the proportion of n-3 relative to n-6 fatty PUFA, with an emphasis on longer-chain PUFA from fish. The Committee on Diet and Health of the Food and Nutrition Board recommended that intakes of n-6 PUFA not exceed the current level, and that individual intakes not exceed 10 en% because information concerning long-term health consequences is lacking. This seems to be prudent advice in view of the potential adverse effects of PUFA in certain diseases.

The *contradiction inherent in any recommendation* to increase consumption of PUFA is that doing so would increase the abundance of lipids that provide the substrate for inflammation and formation of reactive lipid hydroperoxides. Increasing dietary PUFA is not necessary to prevent essential fatty acid deficiency. Figure 11.2 depicts the dilemma in these guidelines as a "phase diagram," in which areas of greatest safety are depicted in terms

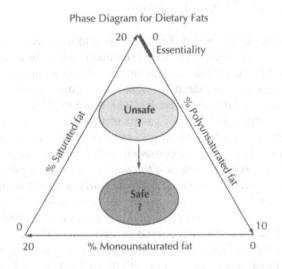

FIGURE 11.2

of percentages of dietary calories contributed by each fat source or nutrient. The NCEP guidelines are equivalent to stating that the compositions indicated in the lower area are safer in terms of reducing risk for heart disease than those in the upper area in the diagram; however, this does not mean that the lower-area compositions are safest for the 60% of Americans do not have coronary artery disease and who might be exposed to more free radicals (note the question marks in the diagram)!

Total fat intake at 5–8 weight% is thought to be totally consistent with the long-term good health of adult monkeys, dogs, rats, hamsters, gerbils, chickens, cows, and catfish. Only cats seem to require diets with higher levels of fat. Adult humans require only 1–2 en% linoleic acid and a trace of linolenic acid (National Research Council, 1989), which can be supplied by one-half tablespoon of corn oil. The single instance in which any public agency has made a recommendation to increase the level of a dietary fat is for polyunsaturated oils, and a recommendation of 10 en% corresponds to 5 times the amount required if heart disease is not present. The National Research Council refrained from making specific recommendations about dietary fat composition in the 1989 Recommended Dietary Allowances. If the aim is to reduce total fat so that energy balance is maintained, and to minimize risks for all-cause mortality including heart disease, cancer, atherosclerosis, stroke, and immunological disorders, then it may be a mistake to recommend consumption of 10% PUFA. There is no biological process that operates at 5–10 times saturation.

11.5 Biological Responses to Polyunsaturated Fatty Acids

Whereas the principal use of dietary saturated fatty acids is to create a long-term energy reserve in adipose tissue, the principal use of long-chain PUFA is for membrane biogenesis. This is most evident if one compares the composition of dietary fat, VLDL particles, and cellular membranes. The VLDL particle typically contains about 53% triacylglycerol and 15% phospholipid, and after circulation and conversion to LDL particles, triacylglycerol is reduced to about 3% whereas phospholipid increases to over 20% by weight (Skipski, 1972). The VLDL/LDL system is designed to transfer principally saturated and monounsaturated fats to adipose and other tissues, with a smaller delivery of unsaturated fatty acids and phospholipids. In addition to the lower rate of transfer of PUFA found in phospholipids, longer-chain PUFA tend to be esterified to C2 in the glycerol backbone, which lipoprotein lipase fails to hydrolyze (Scow, 1977). Fatty acids in this position are thought to be hydrolyzed principally in the liver and oxidized by mitochondrial and peroxisomal beta-oxidation (Havel, 1985).

Now suppose the diet is changed so that polyunsaturated fat is present in equivalent amounts to saturated and monounsaturated fats (as suggested for a guideline of 10:10:10% of dietary calories). The system now makes several adjustments, including (1) proportionally increased delivery of PUFA to adipose; (2) increased synthesis of arachidonic acid and increased content of this fatty acid in membrane phospholipids (Lands et al., 1990; Berdanier, 1992); and (3) substantially increased oxidation of fat with corresponding decrease in deposition in the adipose tissue (Trayhurn, 1992).

11.6 PUFA as Sources for Lipid Signaling Pathways

Although some PUFA are stored in adipose, their normal fate is to become incorporated into phospholipids, which confer fluidity to membrane bilayers. Phospholipases release PUFA from this site and the PUFA are then used to generate lipid mediators. PUFA are precursors for intracellular signaling agents that inhibit lipogenesis in a dose-dependent manner. This inhibitory action occurs in the same range of PUFA found in the U.S. diet (3–10%), and is observed for fatty acids with chain length of 18–24 and at least two double bonds in the 9 and 12 positions (Clarke and Jump, 1996). In this case, n3 and n6 fatty acids are both effective. This inhibition of lipogenesis is a normal response, but indicates a special hormonelike function for PUFA (in the sense that information is being conveyed that reprograms gene transcription for lipogenic enzymes). One cannot assume that actions of these powerful biological mediators are always benign, any more than one could assume that retinoic acid is always safe. Just as cholesterol oxides are probably more potent than cholesterol in regulating gene expression, it is likely that lipid hydroperoxides are more potent than unoxidized PUFA in producing biological responses.

In counterbalance to the inhibitory effect of PUFA on lipogenesis is their action as peroxisome proliferating agents. Peroxisomes metabolize fatty acids by beta-oxidation, and in liver may degrade almost as much acyl CoA as do mitochondria, with the distinction that the reducing equivalents produced in this way are not tightly coupled to ATP synthesis (Nelson, 1992). The ability of PUFA to induce the peroxisomal system may represent a substrate-induced detoxification system. Because long-chain PUFA are potentially toxic, it is essential to control their concentration. Thus, it may be feasible to define nutritional needs for PUFA as lying between the amounts needed for growth and essential functions, and the amounts that will suppress lipogenesis and induce peroxisomal proliferation.

11.7 Basis for the Kinetic Model of Membrane Composition

Lands and colleagues (1990) fit equations to data showing that the PUFA composition of triglycerides and intracellular membranes in rats can be predicted on the basis of dietary fat composition. Moreover, as the en% linoleic acid increased from less than 1% to 4–12%, the amount of arachidonic acid in liver membranes increased from 6% to as much as 30%, with the final composition depending on the amount of competing n-3 and n-9 PUFA. Altering the diet causes a significant change in enrichment of arachidonic acid, and a range of five-fold is feasible with diets that contain 20% corn oil by weight and little competing n-3 fatty acid. Thus it is plausible that eicosanoids produced after activation of phospholipase A2 would be markedly different for diets containing predominantly SAFA as compared to those containing predominantly PUFA. This is one factor that could account for the protective effect of SAFA against tissue degeneration after chronic ethanol exposure (Nanji et al., 1989).

11.8 Use **STELLA®** to Calculate Dietary Fatty Acid Content

Fatty acid composition of common foods can be obtained on the World Wide Web from the U.S. Department of Agriculture's Food and Nutrition Information Service. Table 11.1 gives five common examples of fatty acid compositions for common fats and oils. In general, saturated fatty acids (SAFA) include fatty acids that are 12–18 carbons in length, the major monounsaturated fatty acid (MUFA) is oleic acid (C18:1), the major polyunsaturated fatty acid (PUFA) is linoleic acid (C18:2), and the major n-3 polyunsaturated fatty acid is alpha-linolenic acid (C18:3). Linoleic acid is elongated and desaturated to form arachidonic acid (n-6 C20:4), which is liberated from membranes by the action of an enzyme called phospholipase A2 and converted to proinflammatory mediators that are generically called eicosanoids. Alpha-linolenic acid is elongated and desaturated to form eicosapentaenoic acid (n-3 C20:5) and docosahexaenoic acid (n-3 C22:6), which compete with arachidonic acid for the same enzyme systems and tend to reduce proinflammatory effects. However, all the highly unsaturated membrane lipids can also be oxidized by free radicals, and this problem increases with the larger number of unsaturated bonds. It is crucial to maintain antioxidants such as vitamin E in the cell membranes to mitigate this problem.

Before one can gauge the effects of diet on membrane composition, a way is needed to convert the number of grams of fat in the diet (weight %) to the form used in the equation, which is percent of dietary energy (en %) as each kind of fatty acid.

TABLE 11.1

	SAFA	MUFA	n-6 PUFA	n-3 PUFA
Canola oil	6%	62%	22%	10%
Corn oil	13	25	61	<2
Olive oil	14	77	8	1
Beef fat	52	44	3	1
Margarine	17	49	32	2

11.9 Predict the Effects of Diet on Membrane Phospholipids

Use the model shown in Figure 11.3 and 11.4 to predict the effects of diet on the amount of n-6 and n-3 highly unsaturated fatty acids in cell membranes. Use **STELLA®** to answer the following questions: What is the effect of increasing the total amount of dietary fat over a range of 20% to 40% of calories? What is the effect of changing the sources of dietary n-6 PUFA, such as corn oil and margarine, and what level is effective? What is the effect of varying the levels of saturated fatty acids and monounsaturated fatty acids? If a smaller result is obtained than with an equivalent amount of PUFA, can you determine why on the basis of the equations in the model? Under what circumstances is it desirable to change the amount of PUFA, and when is it not desirable? Is saturated fatty acid always "bad?" Is the ratio of n-3 to n-6 fatty acid important, and why?

Equations Used in the Membrane Model

```
En%_MUFA = 10
En%_n3_PUFA = 2
En%_n6_PUFA = 5
En%_SAFA = 10
MUFA_&_SAFA = En%_MUFA + En%_SAFA
n3_as_%HUFA = 100/(1 + (0.1/En%_n3_PUFA) *
Parameter_for_n3_HUFA)
n6_as_%HUFA = 100/(1 + (0.05/En%_n6_PUFA) *
Parameter_for_n6_PUFA)
Parameter_for_n3_HUFA = (1 + (En%_n6_PUFA/0.05) +
(MUFA_&_SAFA/2) + (En%_n3_PUFA/0.3))
Parameter_for_n6_PUFA = 1 + (En%_n3_PUFA/0.1) +
(MUFA_&_SAFA/2) + (En%_n6_PUFA/0.3)
Tissue_n3:n6_ratio = n3_as_%HUFA/n6_as_%HUFA
Total__En_%_Fat = En%_n3_PUFA + En%_n6_PUFA +
MUFA_&_SAFA
```

11.10 Time Course of Change in Membrane Lipids

The saturation kinetic model provides a basis for estimating changes in tissue lipids, but conclusions should be tempered by two other considerations. First, the equations used here were derived on the basis of experimental studies in rats and the parameters given here predict changes in the liver. Some differences occur among organs, especially the brain, which has specialized

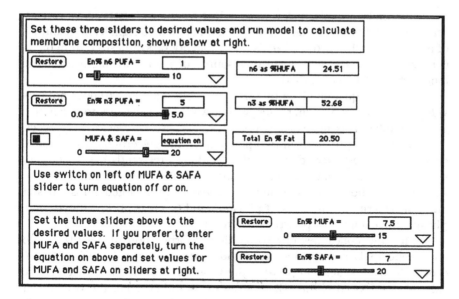

Set these three sliders to desired values and run model to calculate membrane composition, shown below at right.

Restore	En% n6 PUFA = 1	n6 as %HUFA 24.51
	0 — 10 ▽	
Restore	En% n3 PUFA = 5	n3 as %HUFA 52.68
	0.0 — 5.0 ▽	
■	MUFA & SAFA = equation on	Total En % Fat 20.50
	0 — 20 ▽	

Use switch on left of MUFA & SAFA slider to turn equation off or on.

Set the three sliders above to the desired values. If you prefer to enter MUFA and SAFA separately, turn the equation on above and set values for MUFA and SAFA on sliders at right.

Restore En% MUFA = 7.5
0 — 15 ▽

Restore En% SAFA = 7
0 — 20 ▽

FIGURE 11.3

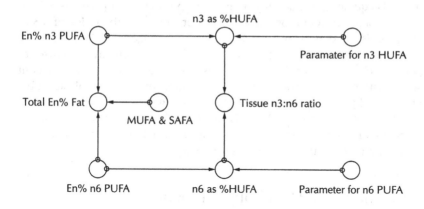

FIGURE 11.4

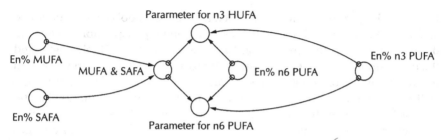

needs for lipids. For instance, it was unclear whether n-3 fatty acids were essential until it became evident that the brain contained significant quantities of n-3, C20:5 and C22:6 fatty acids that could not be synthesized from the n-6 precursor, linoleic acid. Instead, a source of n-3 α-linolenic acid or the longer-chain PUFA is required.

Second, the equation given by Lands and colleagues (1990) does not provide information concerning rates, which would be necessary to generate a kinetic model. This necessity is complex, because fatty acids may be esterified into phospholipids in two positions in the glycerol backbone (sn-1, or sn-2). Saturated fatty acids and oleic acid are preferentially inserted into position 1, whereas arachidonic acid and other PUFA are esterified into the sn-2 position, and the sn-3 position is reserved for the charged, phosphoester unit that determines the class of phospholipid (such as phosphatidylserine, phosphatidylcholine, or phosphatidyl-inositol).

Conner and colleagues (1990) studied incorporation of n-3 PUFA into membrane lipids in rhesus monkeys, and observed that the half-life of C22:6 (docosahexaenoic acid) in phospholipids was 21–29 days, and that a stable value was achieved within 12 weeks. On the other hand, the half-life for arachidonic acid in specific phospholipid classes in specific brain areas is 3 h or less (Shetty et al., 1996). This evidence suggests that the total pool of membrane lipids equilibrates gradually with lipids in lipoprotein particles, which is the important consideration for any attempt to modify the ratios of n-3 to n-6 fatty acids by dietary means. However, exchanges within membrane phospholipid classes that bring about remodeling and facilitate responses to hormones are much faster than this. Let us at least add a kinetic component to the saturation kinetic model on the assumption that the equilibration between dietary lipids and membrane lipids occurs with a half-life of not more than three weeks. Because the brain is separated from the blood by a barrier, it is likely that incorporation into the liver is faster; let us assume the half-life of membrane lipids in the liver equals one week.

11.11 A Model of Phospholipid Kinetics

Figure 11.5 represents a model designed to include a pool of n-6 arachidonic acid and a separate pool of n-3 HUFA in membrane phospholipids. The conditions were set so that the diet initially contained 2 en% linoleic acid and 0.2 en% α-linolenic acid. Rate constants for turnover were arbitrarily assumed to give a half-life of about 7 days for arachidonic acid and 14 days for n-3 HUFA, and synthesis was driven by dietary intake using the saturation kinetic model. By changing the dietary ratio of the n-3 and n-6 essential fatty acids, one achieves a new steady state after several weeks, but significant changes occur within the first week (see Fig. 11.6). It is very likely that this also occurs in practice, because membrane phospholipids are replenished by remodeling, which employs substrates from the nonesterified pool of fatty acids inside cells.

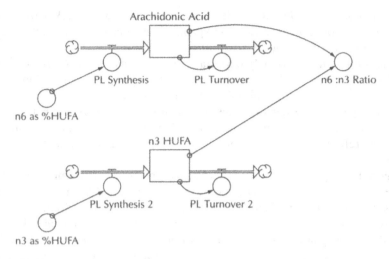

FIGURE 11.5

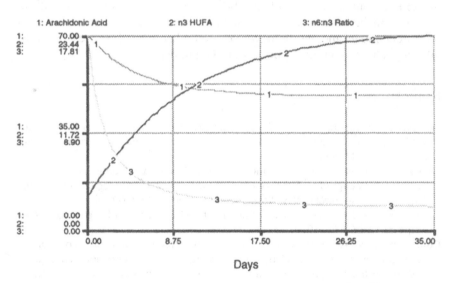

FIGURE 11.6

11.12 Points to Consider

Because phospholipids are comprised of several distinct kinds of lipids (phosphatidylcholine, phosphatidylserine, etc.), and each type of membrane within the cell may have a distinct lipid composition, it is evident that different organs differ significantly in lipid composition and turnover despite being exposed to the same dietary lipids. If one were studying lipid metabolism using isotopic tracers, surely it would be necessary to analyze the data using a multicompartment kinetic model. Yet one might justifiably ask: How much more detail would be necessary to explain to a lay person the potential benefits and risks of changing dietary lipid composition?

This subject has extremely important implications. Just as there is no cell in the body without a functionally crucial phospholipid composition, there is no disease state that is not influenced by this feature, whether it be coronary artery disease, atherosclerosis, kidney disease, arthritis, asthma, cancer, or diabetes. Diets vary enormously worldwide, as do membrane compositions of reference tissues obtained from normal subjects and those with diseases.

Many human studies on fatty acids and disease are epidemiological and have not made use of the saturation kinetic model, and for a variety of reasons, the data are inconclusive and difficult to interpret. Studies on rats clearly show that the levels of different fatty acids can be controlled by diet, and that linoleic and arachidonic acid are involved in tumor promotion during chemical carcinogenesis. This is a clear instance in which a mathematical model could be used to greatly improve study design, and that renders questionable the outcomes of work in which the subjects consumed 30–40% of energy from fats in an uncontrolled manner. Try varying dietary inputs using the kinetic model and determine for yourself what would be necessary to significantly affect membrane phospholipid composition!

References

Berdanier, C.D., B. Johnson, D.K. Hartle, and W. Crowell, 1992. "Lifespan is shortened in BHE/cdb rats fed a diet containing 9% menhaden oil and 1% corn oil." *J. Nutr.* 122 (1992): 1309–1317.

Budowski, P. Alpha-linolenic acid and the metabolism of arachidonic acid. In: Galli, C. and A.P. Simopoulos. "Dietary ω3 and ω6 fatty acids." *Biological Effects and Nutritional Essentiality*. New York: Plenum Press, (1989) 97–110.

Clarke, S.D., and D. Jump. "Polyunsaturated fatty acid regulation of hepatic gene transcription." *Lipids* 31 (1996): S7–S11.

Conner, W.E., M. Neuringer, and D.S. Lin. "Dietary effects on brain fatty acid composition: the reversibility of n-3 fatty acid deficiency and turnover of docosahexaenoic acid in the brain, erythrocytes and plasma of rhesus monkeys." *J. Lipid Res.* 31 (1990): 237–247.

Dyerberg, J., H.O. Bang, E. Stofferson, S. Moncada, and J.R. Vane. "Eicospentaenoic acid and prevention of thrombosis and atherosclerosis?" *Lancet* 2 (1978): 117–119.

Havel, R.J. "The role of the liver in atherosclerosis." *Arteriosclerosis* 5 (1985): 569–590.

Innis, S.M. "Sources of ω3 fatty acids in arctic diets and their effects on red cell and breast milk fatty acids in Canadian Inuit." In: Galli, C., and A.P. Simopoulos. "Dietary ω3 and ω6 fatty acids." *Biological Effects and Nutritional Essentiality*. New York: Plenum Press, 1989, 135–146.

Lands, W.E.M., A. Morris, and B. Libelt. "Quantitative effects of dietary polyunsaturated fats on the composition of fatty acids in rat tissues." *Lipids* 25 (1990): 505–516.

Lands, W.E.M., A. Morris, and B. Libelt. "The function of essential fatty acids." In: Nelson, G.J., ed. *Health Effects of Dietary Fatty Acids*. Champaign, Ill.: American Oil Chemists' Society, 1991, 21-41.

Nanji, A.A., and S.W. French. "Relationship between pork consumption and cirrhosis." *Lancet* 1 (1985): 681-683.

Nanji, A.A., and S.W. French. "Dietary linoleic acid is required for development of experimentally induced alcoholic liver injury." *Life Sciences* 44 (1989): 223-227.

Nanji, A.A., C.L. Mendenhall, and S.W. French. "Beef fat prevents alcoholic liver disease in the rat." *Alcohol Clin. Exp. Res.* 13 (1989): 15-19.

National Research Council. *Nutrient requirements of laboratory animals: rats*. Washington, D.C.: 1978, National Academy of Sciences, 12-18.

National Research Council. *Diet and Health: Implications for Reducing Chronic Disease Risk*. Report of the Committee on Diet and Health, Food and Nutrition Board. Washington, D.C.: National Academy Press, 1989.

Nelson, G.J. Dietary fatty acids and lipid metabolism, in Chow, D.-K., ed., *Fatty Acids in Foods and Their Health Implications*. New York: Marcel Dekker, 1992, 437-471.

Scow, R.O., F.J. Blanchette, and L.C. Smith. "Rate of lipoprotein lipase and capillary endothelium in the clearance of chylomicrons from blood: A model for lipid tranport by lateral diffusion in cell membranes." In Polonovski, J., ed. *Cholesterol Metabolism and Lipolytic Enzymes*. New York: Masson, 1997, 143-164.

Shetty, H.U., Q.R. Smith, K. Washkizaki, S.I. Rapoport, and A.D. Purdon. "Identification of two molecular species of rat brain phophatidylcholine that rapidly incorporate and turn over archidonic acid in vivo." *J. Neurochem* 67 (1996): 1702-1710.

Skipski, V.P. "Lipid composition of lipoproteins in normal and diseased states." In Nelson, G.J., ed. *Blood Lipids and Lipoproteins: Quantitation, Composition, and Metabolism*. New York: Wiley Interscience, 1972, 471-584.

The Expert Panel. "Summary of the second report of the National Cholesterol Education Program (NCEP) on detection, evaluation, and treatment of high blood cholesterol in adults (Adult Treatment Panel II)." *JAMA* 269 (1993): 3015-3023.

Trayhurn, P. "Dietary fatty acids and thermogenesis: Implications for energy balance." In Chow, C.K., ed. *Fatty Acids in Foods and Their Health Implications*. New York, Marcel Dekker, 1992, 517-529.

Weber, P.C. Modification of the arachidonic acid cascade by long-chain ω3 fatty acids. In: Galli, C. and A.P. Simopoulos. "Dietary ω3 and ω6 fatty acids." *Biological Effects and Nutritional Essentiality*. New York: Plenum Press, 1989, 201-211.

12

Responses to Nutrients

There is no thing that is not a poison; it is the dose that makes the poison.

Paracelsus

The nutritionist's credo is "Balance, Variety, and Moderation." This chapter explores a quantitative model that helps explain why caution is always necessary when one contemplates making any drastic changes in the diet to which one is accustomed. There is always some level below which serious deficiencies may occur, and another level above which the likelihood of toxicity or at least interference with other nutrients increases (as indicated in the maxim of the Swiss alchemist, Paracelsus).

Consider the fact that foods containing at least 50 essential nutrients should be consumed in amounts that yield enough energy for daily tasks without any undue excesses or deficits. Plainly, defining one diet that is optimal for human health or for some specific activity is immensely difficult. There is simply too much variability from person to person to be certain that foods are completely balanced, and it is very fortunate that our bodies tolerate variation in the food supply with great resilience.

Despite the complexity of the task, there is a rational basis for defining nutritional needs. In setting the Recommended Dietary Allowances (RDA), the Food and Nutrition Board of the National Academy of Science has employed statistics to define requirements for nutrients (National Research Council, 1989). For individual nutrients, the RDA is calculated based on the average level of need for healthy people, plus two standard deviations of the mean, so that recommendations are sufficiently to supply the needs of 98% of the U.S. population. Information that was considered by the board included the amounts of nutrients needed to maintain rates of growth and development or preserve current body weight, amounts needed to offset obligatory losses and maintain normal reserves, and amounts needed to prevent specific symptoms of deficiency.

Thus, the need for each nutrient must be defined in terms of a function for which it is required. Though there may be no ideal diet, there are certainly diets that support growth and human functions very efficiently, and considerations of threshold, increasing response, maximal response, diminishing returns, and potential for toxicity apply to all nutrients. This chapter will discuss a quantitative theory concerning responses to nutrients.

12.1 Models of Nutrient Response

Responses to nutrients cannot be predicted on the basis of theory without relying on observation and measurement. For energy intake, it is possible to calculate needs on the basis of the metabolic rate and habitual activity patterns. However, for vitamins and minerals, one must determine how much dietary nutrient is required to ensure that important enzymes or binding proteins are at least partly saturated with substrate, and one is never certain that the limiting reaction or binding protein is known. In the absence of this knowledge, it is logical to monitor growth under conditions of deficiency and determine how much nutrient is required to correct the deficiency condition.

It is true that growth curves tend to be S-shaped whether growth is plotted as a function of time or of nutrient level, but all approaches to defining needs require that measurements of food intake and weight gain be made before theoretical curves can be generated. Even though *a priori* models can be constructed (Schulz, 1996), these still require that parameters be estimated by curve-fitting or other mathematical method. Equations that have been used to model nutrient response include a variant of the logistical function (Gahl et al., 1996), saturation kinetics based on analogy with Michaelis-Menten equation (Mercer, 1992), and a general equation called a rational polynomial (Schulz, 1996).

12.2 The Saturation Kinetic Model

Several features of nutrient responses and interactions are well demonstrated by the saturation kinetic model proposed by Mercer (1982 and 1992), which was applied to the ability of individual amino acids to support growth. The food we consume contains protein made up of 20 different amino acids, but our responses to the food depend on the balance of the amino acids found in the protein.

The generalized response curve for any nutrient shows four distinct areas as functions of increasing intake. When intake is between zero and some threshold of need, weight is lost (growth is "negative"). At the threshold, weight is neither gained nor lost, and above that, weight gain increases as intake increases. At higher intakes, less weight is gained as intake increases, and a plateau is attained in which there is decreasing efficiency of gain. Mercer (1992) defined the plateau as the levels of intake that provide at least 95% of the maximum response. Above the 95% response level, less weight is gained. This region of decline may indicate that the nutrient in question interferes with other dietary constituents, or that it becomes toxic above a certain level, which is true of some form of vitamin A, vitamin D, iron, and copper.

$$r = \frac{b(K_{0.5})^n + RmaxI^n + bI^{2n}/(K_s)}{(K_{0.5})^n + I^n + I^{2n}/(K_s)^n} \tag{12.1}$$

In this equation, r = the physiological response that is measured experimentally, I = dietary concentration of nutrient (or % by weight), b = an empirically defined value of r when nutrient intake = 0, Rmax = the maximum theoretical response, n = an apparent kinetic order, $K_{0.5}$ = the half-maximal response for $1/2(R_{max} + b)$, K_s = inhibition constant. The maximum response to the nutrient occurs at a value, $I_{maxr} = (K_{0.5} \times K_s)^{0.5}$.

12.3 Use **STELLA**® to Solve the Saturation Kinetic Equation

A **STELLA**® model was created to solve Eq. 12.1, with the simulation increment used to increase the amount of nutrient in the diet from a value of 0 to a maximum of 10% by weight. The response model is meant to be of a general form that could be applied to humans or any animal; however, it was developed in reference to experiments in which rats were fed diets of defined composition. For a point of reference, consider that human diets typically contain 12–15% protein by weight, and rat diets contain 18–24% protein by weight. Because there are 20 different amino acids, a normal value for each one should be in the range of 0.5–2.0% by weight. The high level map with controls for running the model and the model diagram are shown in Figures 12.1 and 12.2.

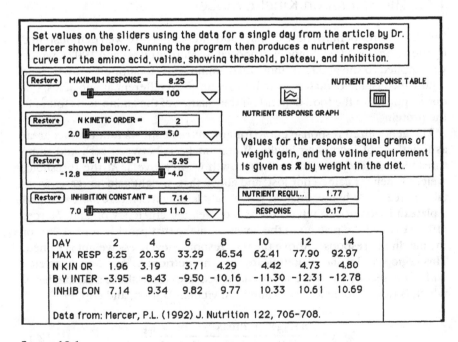

FIGURE 12.1

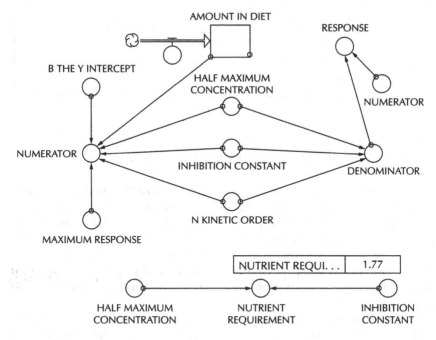

FIGURE 12.2

Equations for the Nutrient Response Model

```
CONCENTRATION(t) = CONCENTRATION(t - dt) +
(CHANGE_IN_CONCENTRATION) * dt
INIT CONCENTRATION = 0
CHANGE_IN_CONCENTRATION = 0.1
B_THE_Y_INTERCEPT = -11.3
DENOMINATOR = (HALF_MAXIMUM_CONCENTRATION^N_KINETIC_
ORDER) + (CONCENTRATION^N_KINETIC_ORDER) +
(((CONCENTRATION^2)^N_KINETIC_ORDER)/
(INHIBITION_CONSTANT^N_KINETIC_ORDER))
HALF_MAXIMUM_CONCENTRATION = 0.44
INHIBITION_CONSTANT = 10.33
MAXIMUM_RESPONSE = 62.41
NUMERATOR = B_THE_Y_INTERCEPT *
(HALF_MAXIMUM_CONCENTRATION^N_KINETIC_ORDER) +
MAXIMUM_RESPONSE * (CONCENTRATION^N_KINETIC_ORDER) +
(B_THE_Y_INTERCEPT * ((CONCENTRATION^2)^N_KINETIC_
ORDER))/(INHIBITION_CONSTANT^N_KINETIC_ORDER)
```

```
NUTRIENT_REQUIREMENT = (HALF_MAXIMUM_CONCENTRATION *
INHIBITION_CONSTANT)^0.5
N_KINETIC_ORDER = 4.42
RESPONSE = NUMERATOR/DENOMINATOR
```

12.4 Running the Nutrient Response Model

The model operates by simulating increasing amounts of amino acid (valine, in this example) from a starting value of 0 to a value of 10% by weight in increments of 0.1%. Because the graph generated by the model shows the response (weight gain in grams) plotted against the simulation increments, the scale on the x-axis runs from 0 to 100. This is the number of increments in the simulation ($100 \times 0.1 = 10$), not the amount of valine in the diet! The amount of valine is viewed in the Table Pad by recording values for the stock called Amount in Diet.

The response of animals (and people) to changes in diet depends not only on diet composition but also on current weight and age. Thus, the parameters needed to solve the equation differ for each day of the experiment. The high-level map (Fig. 12.1) provides a list of four parameters that must be changed to simulated responses on different days. Try adjusting the inputs on the sliders and observe what happens to the magnitude of the response and the shape of the curve.

The graph in Figure 12.3 shows key features of the generalized growth response to nutrients. If the response was a standard Michaelis-Menten curve, it would begin at zero. However, if an essential nutrient is lacking, animals and people do not maintain their preexisting weight—they lose weight. Therefore, it is characteristic to observe that a threshold exists below which weight is lost (zone of loss, left side of graph). Above the threshold, the nutrient is used very efficiently and promotes substantial weight gains (zone of increase). Thereafter, an optimum range or plateau is achieved, which may be defined as growth equal to 95% or more of the maximum. As the concentration of a particular nutrient in the diet continues to increase, it is utilized less efficiently, and may begin to interfere with the use of other nutrients (zone of diminishing returns). This is not necessarily due to a true toxicity, but the result is that the weight gain diminishes and it is wasteful to supply a nutrient at a higher level than needed.

Examples of true interference include effects of diets with imbalanced amino acid content; increased levels of zinc, which may interfere with iron absorption; and increased levels of some dietary fiber interfering with mineral absorption. True toxicities also occur, such as with unduly high intakes of iron, copper, retinoic acid, and vitamin D.

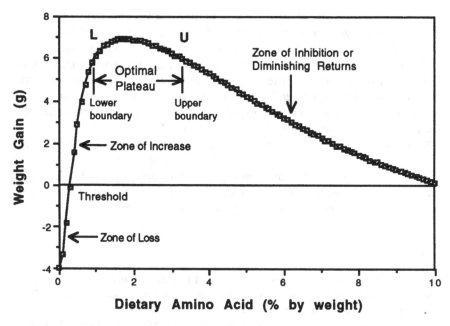

FIGURE 12.3

12.5 Questions and Applications

How could the model be modified to illustrate the concept of *food efficiency*? One way to express food efficiency is the ratio between food intake and weight gain. In what way might this concept be related to the slope of the nutrient-response curve?

As indicated in this chapter, there are several alternative equations for modeling the response to nutrients. An interested student might consider setting up a **STELLA**® model to solve the logistical equation described by Gahl and colleagues (1996) or the polynomial analyzed by Schultz (1996). Each of these equations is supported by an extensive literature and has applications in other areas. Nutritional toxicology is an area of particular importance, and modeling has broad applications in studies of potential toxicity.

12.6 Conclusions

Consider the idea that it is desirable to take in a certain amount of food and food energy; this sets a limit concerning the minimum and maximum amounts of nutrients that will be consumed in a day. Now imagine a series of response curves such as the ones produced by this model, and displayed in Figure 12.3. For each nutrient, there is a region of safety in which all functions will be normal. Below that zone, function will decline and weight may be lost. Above that zone, the nutrients will either be stored, used as an alternative energy source, or may create problems of interference or toxicity.

Mercer (1992) suggested that one way of setting boundaries that define levels of safety is to determine the upper (U) and lower (L) boundaries for the amount of nutrient that gives 95% of the maximum observed response. The ratio of U:L gives an idea concerning margin of safety for any level of intake. It is probably best to attempt to "stay in the zone."

There is at least one source of confusion with this kind of analysis. The test presupposes that all other nutrients are being held constant, and are at least adequate. Although nutritional knowledge has advanced a great deal and one may be safe in making this assumption, nutrients do interact, so that outcomes may differ if other aspects of the diet are changed.

References

Gahl, M.J., T.D. Crenshaw, N.J. Benevenga, and M.D. Finke. "Use of a four parameter logistic equation and parameter sharing to evaluate animal responses to graded levels of nitrogen or amino acids." *Adv. Food Nutr. Res.* 40 (1996): 157–167.

Mercer, P.L. "The quantitative nutrient-response relationship." *J. Nutr.* 112 (1982): 560–566.

Mercer, P.L. "The determination of nutritional requirements: Mathematical modeling of nutrient-response curves." *J. Nutr.* 122 (1992): 706–708.

Mercer, P.L., H.E. May, and S.J. Dodds. "The determination of nutritional requirements in rats. Mathematical modeling of sigmoidal, inhibited nutrient-response curves." *J. Nutr.* 119 (1989): 1465–1471.

National Research Council, Food and Nutrition Board. *Recommended Dietary Allowances*, 10th ed. Washington, D.C.: National Academy Press, 1989.

Schulz, A.R. "Nutrient-response: A 'top-down' approach to metabolic control." *Adv. Food Nutr. Res.* 40 (1996): 227–241.

13

Symmetry of Human Growth and Aging

> [I]t is as though the growth of each organism were regulated by a clock which is wound up at or soon after conception, and then runs down at a constant proportional rate, rather than at a linear rate, as familiar clocks do.
>
> Anna Kane Laird et al., 1965, p. 239

Many of the deepest questions in biology and medicine originate in the processes of growth and development, which are unimaginably intriguing whether considered in terms of complexity, mechanism, or scale. Set aside the question of how the microscopic germ line has been passed along intact over a period of billions of years, yet somehow unfolds to produce creatures that are individually mortal and dispensable. In terms of scale, the human begins with a fertilized ovum weighing a fraction of a milligram, and may attain adulthood at weights of 50–100 kg. My own 15-year-old son, still in his growth years, weighs 113 kg. That represents an increase in mass from conception of much more than one million-fold! It is daunting to consider that we as parents must purchase the groceries to support mass increases on this scale!

However, there are far deeper issues concerning growth than economic ones. A moment's reflection should lead anyone to conclude that an increase in scale of a million or so does not occur on the basis of a linear process, by merely adding an increment to a preexisting size. No, this must stem from compounded growth in which the increase in size depends on multiplication of prior cell mass. This is different from the linear growth that we observe as a child progresses through adolescence. Instead, it is a case in which a major part of the growth depends on the multiplication of cells that have already been produced. This in no way excludes linear growth; we are all familiar with the linear increments in height that accompany the adolescent growth phase. Already, one can see that a model of growth must include at least one linear and one nonlinear component.

Situations in which nonlinear growth predominates include growth of humans and animals from conception to birth, infancy, and childhood; the increase in size of specific organs and tissues (including the fat mass); compensatory growth that occurs after injury to the kidneys or liver; and growth of tumors and cancerous tissues. The growth of populations of humans, plants, and animals—not to mention the replication of *Salmonella* on fried chicken left out at summer picnics!—follows similar laws, so the area of growth has received widespread attention.

In a deeper sense, however, there is a close connection between the rate of growth and the rate of aging, as indicated by Dr. Anna Kane Laird in the quotation beginning this chapter. Once again, the application of a model to a relatively simple process provides evidence for a fundamental law that relates not only to our growth, but also to the passage of time, and indeed, to our demise. And yet, this fundamental property is not at all self-evident, and is widely overlooked. So let us consider some models of growth and then reflect on this fundamental property, for those who wish to somehow extend their lives or "compress mortality" must be aware of this property, or else they are engaged in a fool's quest.

13.1 Simplifying Models of Growth

There are always several means of accomplishing an end. Let us say at the outset that we shall attempt to use the simple stock-and-flow icons of **STELLA**® to describe aspects of human growth, no matter what the prior history of specific equations may be. We are interested in the message, not the medium. We want to see if one model can be produced that can provide realistic answers, and adapted by simple changes to provide a general growth model. It could be a very interesting project for a student to review the major models of growth and attempt to unify them.

There are at least three simple models of growth that have been applied to human development. Each of these models is useful in specific contexts, and there are also compound or multifactorial models in which one seeks to explain growth of each component in a growing creature by applying a basic model to each organ of interest and then summing the terms to simulate growth of the whole individual.

The simplest model of growth is characterized by a single exponential term; it is the model used from the time of Thomas Malthus to the present to explain unlimited growth potential. Any model of multiplicative growth must have an exponential term. However, all growth is limited, and the other models differ in how they treat the processes that limit growth. How do the anabolic processes that build up the organism come into balance with catabolic forces that break it down, or how does cell death eventually come into balance with cell multiplication? Actually, human growth is not quite so simple.

13.2 A General Model of Growth

Let us use **STELLA**® to develop a model that captures different aspects of growth. To keep the model simple, we begin by using one stock, which could represent the developing human body or any of its components. As food and water enter the body, some portion of it may be retained or simply added to the preexisting mass. This would be the process usually called hypertrophy, or growth without addition of new cells. Except for the growing fetus, much growth occurs by cell division, and is proportional to the pre-

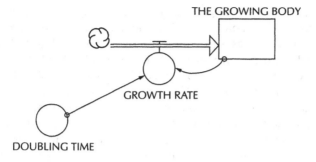

FIGURE 13.1

existing cell mass. Figure 13.1 is a model of simple exponential growth, in which the rate of growth depends upon the number of cells or amount of mass present at any time. Let us state the growth rule followed by this model in English rather than by an equation:

The change in mass of the growing body during a specific interval can be found by multiplying some number (a rate parameter) times the preexisting mass.

The rate parameter has units of time^{-1} and will govern how much time must pass before the mass doubles. Now this representation should look familiar in an important sense, because the parameter multiplied times the existing mass is actually a fractional rate constant (ignore the fact that it does not have to be constant). Recall that the half-life for material in a single compartmental model equaled 0.693 divided by the fractional rate constant. The same is true here, except that mass is increasing instead of decreasing. In other words, we can think of a *doubling time* rather than a half-life. For example, mammalian cells are capable of dividing once a day. If their doubling time is 1 day, then the fractional rate is 0.693/1 d = 0.693 d^{-1}. So let us set up our model of exponential growth to allow the doubling time in days to be entered, and ask **STELLA®** to convert it automatically to the appropriate fractional rate.

A fertilized ovum is about 0.1 mm in diameter, and weighs substantially less than a milligram. Let us set up the model with an initial value of 0.0001 g and a doubling time of 1 day, and see what would happen in a 270-day gestation. Here are the equations used:

```
THE_GROWING_BODY(t)  =  THE_GROWING_BODY(t - dt) +
(GROWTH_RATE)  * dt
INIT THE_GROWING_BODY = 0.0001
INFLOWS:
GROWTH_RATE = (0.693/DOUBLING_TIME) * THE_GROWING_BODY
DOUBLING_TIME = 1
Weight_kg = THE_GROWING_BODY/1000
```

With this assumption, our growing fetus reaches 21 kg in 30 days, and our mother would not be able to eat fast enough to keep up thereafter! One might ask if there are cases in which strictly exponential growth is used as part of models, and the answer is yes. For example, Bonds and colleagues (1984) observed that accumulation of fetal fat could be predicted with an un-inhibited exponential model, with an amount of fat equal to 0.0039 g at 182 days of gestation and a rate of increase of 0.00429 per day, which yielded about 650 g at the time of birth. How interesting that even before birth fat seems to be characterized by exuberant accumulation! However, the simple model is mostly useful to illustrate the concept of doubling time and demon-strate conclusively that this is not what happens in normal growth.

13.3 A Model with a Declining Growth Rate

In a growing fetus, there are several reasons why unrestrained growth does not occur. Once cells begin to differentiate and form organs and tissues, they generally cease dividing. In terms of cell cycle kinetics, they move out of the cycle and into the non-dividing stage called G_0. In some cases, the differenti-ated cell simply can not divide; in other cases, it may be true that a growth inhibitor has been produced, or that a growth stimulator is being produced at a lower rate.

One way to model this is to reduce the growth rate in a manner that is re-lated to the mass of the cells, so that the growth rate declines until it reaches zero when a mass limit is attained. Bonds and colleagues (1984) modeled the growth of human infants with this growth model, using the following pa-rameters: Observed weight at the 26th week of gestation, 340 g; weight at the end of gestation, 3,731 g; rate of growth per day, $9.0 \times 10^{-6} \, (g \times d)^{-1}$. The **STELLA®** model (Fig. 13.2), equations, and outcome (Fig. 13.3) are shown next.

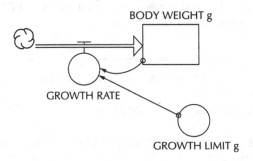

BODY WEIGHT g

GROWTH RATE

GROWTH LIMIT g

FIGURE 13.2

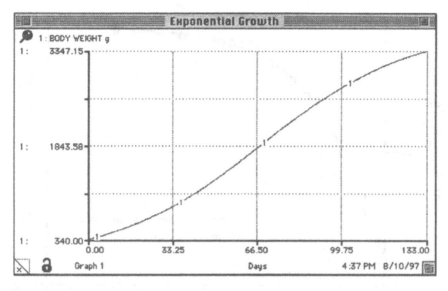

FIGURE 13.3

Equations

```
BODY_WEIGHT_g(t) = BODY_WEIGHT_g(t - dt) +
(GROWTH_RATE) * dt
INIT BODY_WEIGHT_g = 340
INFLOWS:
GROWTH_RATE = 0.000009 * BODY_WEIGHT_g *
(GROWTH_LIMIT_g - BODY_WEIGHT_g)
GROWTH_LIMIT_g = 3731
```

Now, instead of setting the growth rate to a constant, observe that the growth rate is found by multiplying the fractional growth rate by body weight by *the difference between the limiting body weight and the current body weight*. This is one form of the *logistic equation*.

This very simple model has other applications. For instance, the liver is one of the few organs that will regenerate in the human adult after it is injured. Frequently, part of the liver must be removed in the treatment of trauma or carcinoma, and the liver regrows until it is close in size to the original mass. Though the mechanism is not clear, the growth involves increased cell division in response to newly produced hepatocyte growth factor; whether there is also a specific growth inhibitor is still under study. Chen

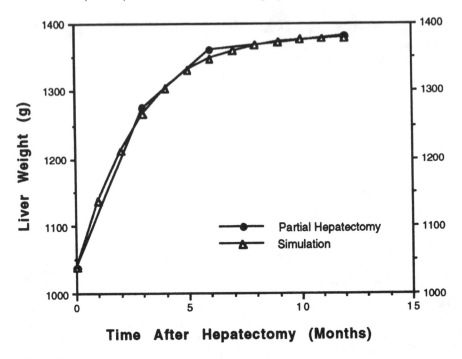

FIGURE 13.4

and colleagues (1991) studied the rate of hepatic regrowth after hepatectomy, and the parameters in the limited growth model were changed to simulate their results, as shown in Figure 13.4.

13.4 Specific Growth Rate

Now let us modify the limited growth model to extract one more piece of information. The target that we have set for the model is the average birth weight of babies in a particular city, 3.73 kg (8.2 pounds) within a range of 3.5–4.0 kg. This is not intended to include growth during infancy. Nevertheless, it is instructive to plot a value called the specific growth rate, defined as the change in growth increment divided by the body mass already attained. This is plotted in Figure 13.5 for the same data on infant growth.

The major outcome of human growth is that the multiplication of cells eventually nearly ceases, and cells differentiate and become nondividing. The cells are not dead—they have taken on mature functions, which removes them from the pool of dividing cells. The specific growth rate shows this very clearly; as time progresses, there is less and less capacity for renewed growth. Our infant has barely been born, yet its growth potential is limited! And that is true; for instance, all the nephrons it will ever have are already present,

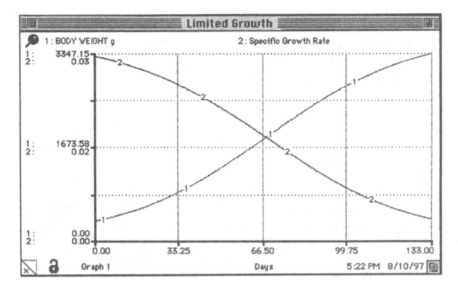

FIGURE 13.5

and if these are lost, there is no means except kidney transplantation to obtain more.

Whereas some stem cells persist for very long periods, differentiated cells have finite lifespans, which may be foreshortened by injury, lack of oxygen, or exposure to pathogens or toxins. Consider for a moment the outcome of these two facts: Cells have finite lifespan, and the reserve capacity for regenerating new cells or for repairing tissues declines with age. Fortunately, the picture is not bleak for our newborn infants, for they will continue some growth by cell division, and their linear growth rate will come to predominate. Strictly speaking, linear growth would involve a process that would be modeled as a constant rate, not a constant proportion.

13.5 The Gompertz Growth Function

A different method of modeling inhibited growth was named for the actuary, Benjamin Gompertz (1825), who devised a method related to changes in mortality rates during the lifespan. The equation can be modified so it demonstrates a form of limited growth. Basically, the Gompertz growth function could be related to change in weight as follows:

$$\frac{dW(t)}{dt} = \gamma W(t)$$

$$\frac{d\gamma}{dt} = -\alpha\gamma$$

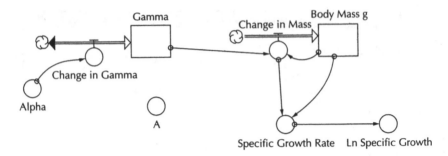

FIGURE 13.6

where W(t) is the weight at any time, γ is a proportionality constant that is also time dependent, and α is a constant that shows how γ changes with time. In this equation, α is the factor that determines the time scale of the process of growth. The solution to the equation is a complex exponential function (Laird et al., 1965):

$$W(t) = W_0 \cdot Exp\left\{\frac{A}{\alpha}\left[1 - Exp(-\alpha t)\right]\right\}$$

In this equation, A is the value of γ at t_0. Let us use **STELLA**® to solve the equation and observe how it behaves.

Equations

```
Body_Mass_g(t) = Body_Mass_g(t - ) + (Change_in_Mass)
* dt
INIT Body_Mass_g = 0.062
INFLOWS:
Change_in_Mass = Gamma * Body_Mass_g
Gamma(t) = Gamma(t - dt) + (Change_in_Gamma) * dt
INIT Gamma = A
INFLOWS:
Change_in_Gamma = Alpha * Gamma
A = 0.267
Alpha = -0.0304
Ln_Specific_Growth = LOGN(Specific_Growth_Rate)
Specific_Growth_Rate = Change_in_Mass/Body_Mass_g
```

A plot of the results with these initial values, which are appropriate for growth of a guinea pig (Laird, 1965), is shown in Figure 13.7.

Observe that the Gompertz function also produces a sigmoidal growth curve. The rate of accumulation of body mass depends on a coefficient, γ,

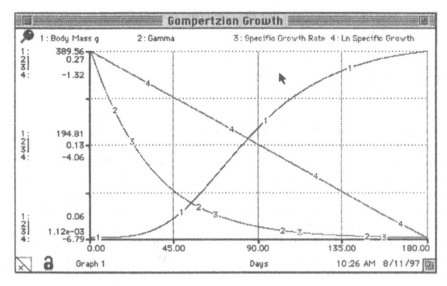

FIGURE 13.7

that is high initially but that decays in simple exponential fashion (line 2 in the graph). The specific growth rate, or change in mass per unit time divided by the preexisting mass, declines in the same, simple exponential fashion. If this is plotted on a logarithmic scale (line 4), a straight line is produced whose slope is the rate at which growth declines.

13.6 The Relationship with Aging

If one compares the outcomes of the logistic growth function with the Gompertzian function, one difference is that the Gompertzian function indicates that the specific growth rate (presumably an index of the ability of tissues to replenish through mitosis) declines exponentially throughout life. In Figure 13.7 the first data point is at 20 days after conception, and birth occurs at about 70 days after conception. Thus, the specific growth rate continues to decline well into adult life for the guinea pig, and the same is true for all animals. The rate will asymptotically approach zero, meaning that the ability to restore tissues after injury becomes very limited as cellularity declines.

All cells have finite life spans. For this reason, one could include in the models just discussed a small loss of mass that increases over time. Indeed, loss of lean body mass and body water occurs at a modest rate as we age. This leads to an accumulation of extracellular matrix that is also aging because the fewer cells embedded in the matrix have limited capacity to repair any damage or add new elastic components. How striking a realization it is that our specific growth rate is declining before we are born, indeed, it must lest we become giants—and this decline is inexorably linked to senescence and the aging process!

13.7 The Symmetric Relationship Between Relative Growth and Physiologic Time

A singular insight occurred to Samuel Brody (1937) concerning the relativity of growth rates, physiological time, and physiological weight. His paper is often cited as one of the first observations that metabolic rate should be related to a power function of body mass. His equation is: metabolic rate, Q, equals $70.5*M^{0.734}$, where mass is in kg and the factor is kilocalories per kg. Professor Kleiber extended and attempted to explain Brody's work in terms of theory. However, Brody's major reason for writing the paper was not to restate a conclusion he had published earlier in the Missouri Agricultural Experiment Station Research Bulletin, but rather to indicate that *the half-time of an animal's growth rate is a potentially useful standard for physiological time.* His logic was that if one plots the growth of any animal or plant as the currently attained fraction of its adult weight, the curves are superimposable. He suggested that if one extrapolates this curve to a weight of zero, a growth rate constant can be obtained that differs for every animal. His results also showed that growth curves are not monotonic, because growth during the fetal-juvenile stages cannot be matched by the simple model. Humans, in particular, have an incredibly protracted juvenile growth phase, and do not match the simple asymptotic curve until the age of about 15.

This intrinsic growth rate defines the time required for a human or animal to attain half of its remaining growth. The rapidly growing rat has a rate constant, k, equal to 0.644 yr^{-1}; the cow has a rate constant of 0.054, so that the rat approaches maturity almost 12 times faster than a cow. Thus rodents are able to multiply far faster than the large herbivores or the scientists who observe them. Dr. Brody then indicated that, just as metabolic rate could be scaled to $Mass^{0.734}$, physiological time could be scaled as e^{kt}. It is also feasible to plot senescence in this manner, which produces a very interesting symmetry between the phase of life when one is growing and the phase during which one ages. The growth function is an asymptotically increasing exponential function; the senescence function asymptotically declines. Here, one may observe that inasmuch as one never attains the asymptote, there is never any surety about one's demise!

13.8 The Brody Model for Growth and Senescence

During an extensive summary of growth among different animals, Brody (1927) devised a time scale for growth based on a simple asymptotic increase in mass. He extrapolated the growth curve that is observed between the postpartum period and maturation back to a time when weight would equal zero

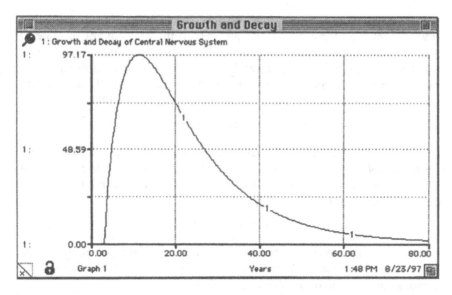

FIGURE 13.8

if growth followed a simple exponential function with a rate parameter of k per year. He suggested on this basis that, at least during these years, one way of expressing the relative time scale would be to define a physiological time unit equal to $k(t - t^*)$, where t is the present age of the animal or human, t^* is the extrapolated time when weight equals zero, and k is the parameter that characterizes the rate of growth. Brody also noted that it is possible to treat growth and senescence (loss of any specific physiological capacity) as the sum of two such functions that have different rate constants. Stated as an equation, his method could be written as follows:

$$\text{Function of interest} = C^* e^{-kt} - D^* e^{-k't}$$

where the function may be body mass, vitality, reciprocal of a death rate, or any other index. STELLA® was used to solve this equation for growth and decline of the central nervous system, using the equation, $V = 331(e^{-0.07t} - 1.37e^{-0.19t})$ (Brody, 1927). The curve is shown in Figure 13.8. The data indicate that the rate at which the human nervous system loses function is about 7% per year. What is of interest is the concept of the symmetry between growth and loss of function, admitting that the mechanisms that underlie each are probably different.

13.9 A Time Unit Based on Growth: The Chron

Augustus J. Fabens (1965) greatly extended Brody's observation about the relativity of time in a very interesting presentation of another growth model devised by Ludwig von Bertalanffy (renowned as a founder of general system dynamics).

The von Bertalanffy (1938) model was derived from the consideration that growth is related to the metabolic rate, which therefore ought to scale to Mass$^{2/3}$. If one agrees that the Brody-Kleiber relationship is closer, the empirical description would become Mass$^{3/4}$. However, the rate of tissue breakdown is generally a first-order function of body mass (k*M). Therefore, the growth function for weight ought to fit a model written as follows:

$$dM = k*M^{2/3} - k'*M$$

where dM is the change in body mass and k is an *anabolic* rate parameter and k' is a *catabolic* rate parameter. The outcome of this model differs from that of Brody's model, which actually fits the change in a linear dimension (call it x) such as height or length, as follows:

$$x = a(1 - be^{-kt})$$

where a, b, and k are constants that depend upon the species of animal under consideration. In contrast, the von Bertalanffy model describes the change in mass (related to volume), as follows:

$$M = [a(1 - be^{-kt})]^3$$

What is interesting about this curve for growth in mass is that it is similar to the simple asymptotic one for length, but has an inflection near the beginning just as Dr. Brody's data clearly demonstrated. There is an inflection in the empirical data, and one may be able to justify this as the time at which exponential fetal growth ceases, and linear growth toward adult proportions begins.

13.10 Time Standards for Growth and Aging

Nabokov, the notorious punster, wrote, "Time is but a quack in the court of Chronos." At about the same time, Fabens (1965, p. 276) wrote: "Just as the half-life is the natural unit for describing the age of a sample of a radioactive element, the period ln2/k, in which the animal achieves half its remaining linear growth, is a natural unit for the age of an animal . . . I suggest that this period be called a *chron* after the familiar Greek word, ΧρονοȢ, 'time.'"

Time in chrons = Ordinary time*(k/ln2)

This is fascinating; we now have a definition of "ordinary time!" From the von Bertalanffy model, one immediately extracts three important conclusions. First, it is likely that fetal growth must be considered separately from growth during the period from infancy through adolescence and even our approach to The Asymptote of death. Second, a natural time base for the age of animals emerges; ln2/k, where k is a species-specific rate constant. *Because it depends on mass, the time unit also depends on metabolic rate!* Third, there is a separate function that describes tissue breakdown; in this model, growth functions are characterized by a different rate parameter than catabolic processes. The question then emerges: is senescence more closely related to growth or tissue breakdown? Or is it some sort of chimera?

It seems that in considering von Bertalanffy's basic model, which is grounded in theoretical principles, we have unwittingly arrived at a profound mystery. And so let us leave it a mystery; it is worth mulling, and worth experimentation. With some humility, one recalls that it was 60 years ago that Samuel Brody wrote: "The purpose of the present paper is to indicate a *practical* application to growth in particular and biology in general of the idea of relativity of time and weight."

The value of computer modeling is also that it leads one to explore unfamiliar terrain. I had never heard of Fabens' work or the von Bertalanffy model of growth before attempting this chapter. I recommend both as stream beds that may yet return nuggets of truth to individuals who are willing rework this claim and move between simulation and experimentation. In what ways have these models been extended in subsequent years, and to what extent have experiments tested their premises? We are all encouraged (or warned) to look into this before *our* relative times equal ln2/k!

13.11 Summary and Applications

This work suggests that there are at least three related ways of thinking about growth and physiological time. The Brody model scales physiological time to a factor, k(t − t*); using this curve allows growth of many different animals to be plotted on the same relative scale. This permits one to note exactly how many rat-months equal one human-year (Brody, 1937). Laird and colleagues (1965) consider the rate at which an animal's specific growth rate declines, and similarly suggest that this loss of regenerative capacity can be described as an exponential function with a rate constant. Fabens (1965) derived a similar value from the von Bertalanffy growth model and proposed a definition for a unit of physiological time he called a *chron*.

The basic models shown here could easily be made more comprehensive. For instance, growth occurs by cell multiplication and by increases in length and accretion of extracellular material. Koops (1986) has described growth in terms of several linked logistic functions, and Laird (1964) examined the relative growth of the human embryo in terms of growth of several individual organs. Bonds and colleagues (1984) obtained excellent simulation of human fetal growth by adding two logistic functions (lean tissue plus body water) and one exponential term for increased fat mass.

A related area of extreme interest is modeling the growth of tumors and the effects of therapy on tumor growth. Laird (1964) and Norton and Simon (1977) used the Gompertz function to model tumor growth and the effects of radiotherapy on growth rates. Students interested in effects of cancer and outcomes of therapy will find an extensive literature that could be explored profitably using a simulation program.

And last, an enthusiastic student may be repaid by using **STELLA**® to create a von Bertalanffy growth model, and then using it to make inferences about growth and senescence.

References

Bonds, D. R., B. Mwape, S. Kumar, and S.G. Gabbe. "Human fetal weight and placental weight growth curves. A mathematical analysis from a population at sea level." *Biol. Neonate* 45 (1984): 261–274.

Brody, Samuel. "Growth and development with special reference to domestic animals. X. The relation between the course of growth and the course of senescence with special reference to age changes in milk secretion." *Missouri Agr. Exp. Sta. Bull.* 105 (1927): 4–64.

Brody, Samuel. "Relativity of physiologic time and physiologic weight." *Growth* 1 (1937): 60–67.

Chen, M.-F., T.-L. Hwang, and C.-F. Hung. "Human liver regeneration after major hepatectomy. A study of liver volume by computed tomography." *Ann. Surg.* 216 (1991): 227–229.

Fabens, Augustus J. "Properties and fitting of the von Bertalanffy growth curve." *Growth* 29 (1965): 265–289.

Gompertz, B. "On the nature of the function expressive of the law of human mortality, and on the new mode of determining the value of life contingencies." *Phil. Trans. R. Soc. London* 115 (1825): 513–585.

Koops, W.J. "Multiphasic growth curve analysis." *Growth* 50 (1986): 169–177.

Laird, A.K., "Dynamics of tumor growth." *Br. J. Cancer* 18 (1964): 490–502.

Laird, A.K., "Dynamics of relative growth." *Growth* 29 (1965): 249–263.

Laird, A.K., S. Tyler, and A.D. Barton. "Dynamics of normal growth." *Growth* 29 (1965): 233–248.

Norton, L., and R. Simon. "Growth curve of an experimental solid tumor following radiotherapy." *J. Natl. Cancer Inst.* 58 (1977): 1735–1741.

Wheatley, J.M., N.S. Rosenfield, L. Berger, and M.P. LaQuaglia. "Liver regeneration in children after major hepatectomy for malignancy—Evaluation using a computer-aided technique of volume measurement." *J. Surg. Res.* 61 (1996): 183–189.

14

A Stochastic Model of Senescence and Demise

[I] should prefer to any ordinary death, being immersed in a cask of Madeira wine, with a few friends, until that time, to be then recalled to life by the solar warmth of my dear country.

Benjamin Franklin

The Centers for Disease Control and Prevention publishes a weekly periodical cheerfully named *Morbidity and Mortality*, wherein one may discover the latest causes of sickness and death in the United States. Those who dutifully read this government periodical do so, no doubt, more in the hope of avoiding the dire outcomes it describes than in learning why funeral attendance may be up or down lately. However, as the old syllogism explains, all men are mortal (a little more so than women, it seems). And as one might imagine, since the time of Gompertz, several models of risks for death have been devised. One of the best sources of such models and a great deal of other information about aging is Bernard Strehler's classic book, *Time, Cells, and Aging*. The model developed in this chapter does not explicitly duplicate any of the six discussed in that volume, but captures the flavor of decline in function coupled with an element of randomness that tests the ability of the human system to withstand life-threatening stresses and injuries.

14.1 Decline of Function with Aging

The symmetry between growth and aging suggests a number of possible means whereby the decline in physiological capacities and the increasing risk for disease and death could be modeled. The simplest is to begin by describing a decline in physiological function over time. In principle, the risk for death must be related to the ability of each of our vital organ systems to repair itself after a health-threatening event. The course taken in the model shown next was to add a functional capacity and a reserve capacity that increase in magnitude until the age of 25, and to offset this by a loss of function that is small at the outset but that increases rapidly after the age of 25. Parameters were adjusted so that the capacity approaches a limit of zero at the age of 115.

The idea of elasticity was added by calculating a survival limit that is much lower than the total capacity, but that asymptotically approaches a value of zero. Any stressor that exceeds this limit is likely to cause death; any stressor that does not exceed the limit will allow recuperation.

14.2 An Element of Randomness

We all recognize that many life events occur at random. Some of these may be age-independent; it may be as likely that one suffers a serious accident at 20 years of age as it is at 60. On the other hand, some events clearly depend on age, yet are also unpredictable; an example may be exposure to a pathogenic microorganism that becomes more infectious as our immune function declines. If one wished to gain an accurate picture based on population statistics, it would be necessary to add together all the risks that apply throughout life. Let us instead take an arbitrary approach and set values for age-dependent and age-independent risk factors that yield a finite chance of dying at any age, but that increases so that the likelihood an individual will expire prior to the age of 100 years is very high. This was done by using these two kinds of risk as input variables for a Monte Carlo number generator. This device produces a number between 0 and 1.0, in this case with a frequency that depends on the ratio between the two risk factors and the remaining capacity of the physiological function under consideration. The output only indicates that a life-threatening event has occurred, not its magnitude relative to our subject's capacity.

The model was created under the assumption that chance acts on our gradually declining system in two ways. First, there is no way to predict when an individual will be exposed to a life-threatening change. Second, because one does not know the day on which an event will occur, neither does one know whether one's ability to deal with the stress will be optimal or reduced. For instance, there may be a high likelihood that the health of people enduring a famine or living in a war zone may be jeopardized before a specific injury or threat occurs. Therefore, the model was created so that if a life-threatening event occurred, a second random number generator was used to predict the magnitude of the threat and compare it to the reserve capacity. Death was assumed to result from such an injury or event if it exceeded the survival limit. Figure 14.1 shows the diagram that was generated to produce this model.

Equations for the Model of Senescence and Risk for Death

```
The_End(t) = The_End(t - dt) + (End_Game) * dt
INIT The_End = 0
INFLOWS:
End_Game = Demise * 4
Age_Dependent_Factor = 4
Age_Independent_Risk = 100
Demise = IF(Risk_Level > Survival_Limit) THEN(1)
ELSE(0)
```

```
Function = 40 + 40 * (1 - EXP(-Growth_Rate * TIME))
Growth_Rate = 0.14
Relative_Capacity = (Total_Capacity/87.67) * 100
Reserve = 10 + 10 * (1 - EXP(-Reserve_Rate * TIME))
- 10 * (1 - EXP(-0.04 * TIME))
Reserve_Rate = 0.139
Risk_Factor = 0.6
Risk_Level = IF(Time_and_Chance = 1)
THEN(RANDOM(0,Risk_Factor * Relative_Capacity)) ELSE(0)
Senescence = IF(TIME> = 25) THEN(90 * (1 - EXP(-
Senescence_Rate * (TIME - 25)))) ELSE(0)
Senescence_Rate = 0.018
Survival_Limit = IF(Relative_Capacity> = 0) THEN(0.45 *
Relative_Capacity) ELSE(0)
Time_and_Chance = IF(The_End < 1)
THEN(MONTECARLO((2 * Age_Independent_Risk +
Age_Dependent_Factor * TIME)/Relative_Capacity))
ELSE(0)
Total_Capacity = Function + Reserve - Senescence - 0.1
* TIME-10 * (1 - EXP(-0.01 * TIME))
```

Figure 14.2 demonstrates that the function of interest increases from the value it takes at birth to a higher level at maturity, and then begins to decline. The shape of this curve will differ for every physiological function one might consider, whether strength, a mental ability, kidney function, lung function, immunological reserve, and so on. What is interesting about a stochastic model such as this one is that the outcome will never be the same twice. The pattern of life-threatening events and the response to them are both gener-

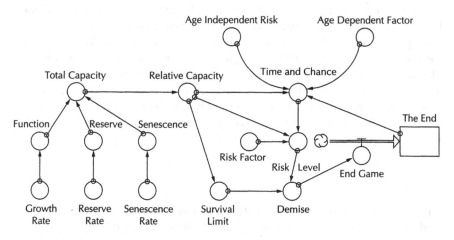

FIGURE 14.1

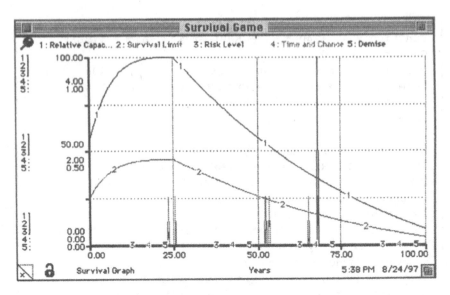

FIGURE 14.2

ated randomly. Based on the structure of the model, one can predict that there is a higher likelihood that subjects will die as age advances, but on any trial, the subject may die in any decade. The graph demonstrates that the subject shown withstood perhaps six life-threatening events before one exceeded his survival limit at the age of about 68. Another person may experience none until the age of 80; yet another may suffer multiple events in childhood. With this model, **STELLA**® is playing dice with the universe.

14.3 Applications

There are many models that incorporate declining physical capacity with various factors that increase risk. A student interested in this area might consider what parameters could be changed that would enhance longevity, and what the real-world correlates of these might be. It would be easy to add a particular factor that increases risk, such as frequency of cigarette smoking, drug abuse, or risky sexual behavior. Better yet, locate a model in the epidemiological literature and reproduce it using **STELLA**®. For inspiration, be sure to consult *Morbidity and Mortality Weekly*! It is available on the Internet at the CDC home page.

References

Hayflick, Leonard. *How and Why We Age*. New York: Ballantine Books, 1994.
Strehler, Bernard L. *Time, Cells, and Aging*. New York: Academic Press, 1977.

15

Mortality and Risk for Chronic Disease

To think of a man without duration is as ridiculous as thinking of him without an inside.

Sir Arthur Eddington

Most people are interested in the positive aspect of taking risks. We accept risks to maximize successes, health, fitness, and happiness. We prefer to look at the sunshine and not the shadow, and early in our lives, that is usually the dominant theme, though that has not always been the case in human history. The concept of health risk in relation to diet and lifestyle may be unclear in the absence of a means for quantification. In a statistical sense, risk is defined as the likelihood that an individual will die or suffer a specific disease during a stated interval. Although risk cannot truly be determined for a given individual, levels of risk associated with particular behaviors or conditions can be estimated from epidemiological correlations between the factor of interest and the incidence of disease or death in a population. It is possible to estimate accurately the number of people who will die from cancer of the colon next year in the United States, but utterly impossible to say who will die. The goals of this chapter are to discuss general ideas about risk and to use **STELLA®** to simulate the risk for coronary artery disease (CAD). The model should quantitatively describe the relationship of the specific behaviors to predicted life span. However, we are justified in emphasizing the positive side of this—namely, risk reduction, prevention, and the advantages of survivorship. One would like to know what preventative steps to take early, and not merely what palliatives may be available when a disease has irreversibly progressed.

15.1 Morbidity and Mortality

The leading causes of death in the United States are heart disease, cancer, injuries, stroke, chronic lung disease, pneumonia/influenza, suicide, diabetes, liver disease, atherosclerosis, and HIV infection. Several major health risk factors are associated with the first five major causes of death. Each risk factor independently contributes to aggregate risk and the risk associated with any one factor is compounded by the presence of the others; this is the concept of *comorbidity*. Living habits, obesity, and diet affect the level of risk for these diseases, and behavioral changes have saved many lives by reducing these risk factors (Kannel, 1990; Sytkowski et al., 1990). Many of the leading causes of death are preventable wholly or partially through changes in lifestyle as individuals assume personal responsibility for their health. (MRFIT Research Group, 1982 and 1990; USDHHS, 1990; Kannel, 1990 and 1993).

Epidemiologic, clinical, and laboratory evidence indicates that diet influences the risk of several major chronic diseases. The evidence is very strong for atherosclerotic cardiovascular diseases and hypertension and implicates diet in certain forms of cancer. Certain dietary patterns predispose to dental caries and chronic liver disease, and positive energy balance produces obesity and increases the risk of noninsulin-dependent diabetes mellitus. The evidence is not sufficient to determine if dietary patterns influence osteoporosis and chronic renal disease. When evidence is clear, dietary modification can reduce known risks, and it is important to inform the public about known risks and the possible benefits of dietary modification (National Research Council, 1989; USDHHS, 1988). Although many dietary components are involved in the relationship between diet and health, a disproportionate consumption of foods high in fats is a major factor, especially when it reduces intake of foods high in complex carbohydrates and dietary fiber that may be more conducive to health (USDHHS, 1990).

15.2 Ideas about Risk

Before discussing risk reduction, it is important to gain insight into ways that risk is defined. In his book, *Time, Cells, and Aging,* Bernard Strehler (1977) elegantly summarized the development of quantitative theories about aging and risk. It was Benjamin Gompertz who in 1825 observed that mortality rates in many animals increase exponentially. Essentially, a mortality rate can be viewed as the likelihood that a specific number of individuals from a population of a specified size will die during a stated interval. The rate can never tell which individuals will succumb, but it does provide a statistical measure of expectation. The Gompertz equation for mortality rate, R_m, is:

$$R_m = -\frac{1}{n} \times \frac{dn}{dt} = R_0 e^{\alpha t} \tag{15.1}$$

in which n = number of individuals at time t and α and R_0 are constants, α is a mortality rate constant, and R_0 is the hypothetical mortality rate at t = 0. This relationship produces an exponentially increasing mortality rate as age increases. This relationship was modified by adding an age-independent term, A, so that

$$R_m = R_0 e^{\alpha t} + A \tag{15.2}$$

For instance, the risk of having a particular kind of accident may be constant throughout life. This provides an equation for an average mortality rate; in a moment, we will consider how this relates to ideas about risk for a particular disease and ideas about risk reduction. First, however, let us find out what Eq. 15.2 predicts.

15.3 Use **STELLA**® to Solve the Gompertz Equation

A model created to solve the Gompertz equation is shown in Fig. 15.1. An initial population of 100,000 individuals was entered into the stock labeled Remaining Population. In this case, the mortality rate was assumed to include an arbitrary age-independent rate of 10^{-6} per year and an age-dependent rate. The age-dependent rate, R_0, was assumed to have an initial value equal 6.5×10^{-4} per year, which must be multiplied times an exponential function, $e^{\alpha t}$. In the diagram, α was designated "Change over time."

Mortality Model Equations

```
Remaining_Population(t) = Remaining_Population(t - dt)
+ (-DEATHS_PER_YEAR) * dt
INIT Remaining_Population = 100000
OUTFLOWS:
DEATHS_PER_YEAR = Remaining_Population * MORTALITY_RATE
CHANGE_OVER_TIME = 0.05
EXPONENTIAL_FUNCTION = EXP(CHANGE_OVER_TIME * TIME)
LIMITING_RATE = 0.00065
MORTALITY_RATE = 10^-6 + LIMITING_RATE *
EXPONENTIAL_FUNCTION
```

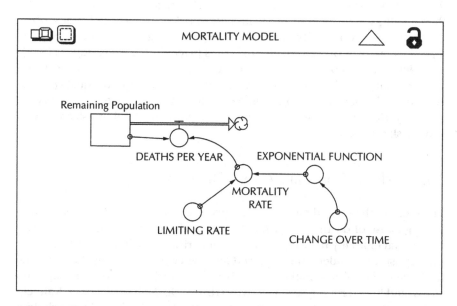

FIGURE 15.1

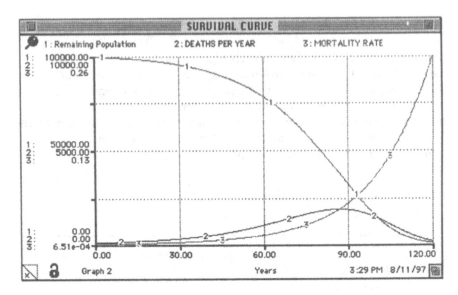

FIGURE 15.2

Figure 15.2 represents survivorship in a population of 100,000 people based on a solution to the Gompertz equation for mortality due to a single cause. Note that initially the death rate is very low, but it increases exponentially over time, so that half the population has died by the age of 80, and most are dead by the age of 120 years. More accurately, one could write a similar function for each cause of mortality that increased with age, and add to that any risk for death that increased in any other fashion, such as a constant risk for accidents or exposure to a pathogen. The aggregate death rate would then result from the sum of all the individual risks. Strehler (1977) describes other theories of mortality and ways of representing the decline in physiological capacity or overall vitality that takes place during aging. Riggs (1992) provides an excellent summary of these concepts and tables giving mortality data, including coefficients that can be used in the Gompertz equation, and Strehler and Mildvan (1960) provide a useful introduction to the theory of mortality. However, let us now examine one particular kind of risk and ideas about risk reduction.

15.4 Risk for Coronary Artery Disease

The focus of the second model is coronary artery disease (CAD), which is the leading cause of death in the United States and accounts for nearly one-half of the total mortality rate. The nonmodifiable risk factors for CAD are genetic predisposition, gender, and advanced age, whereas lifestyle-related risk factors include cigarette smoking, excessive body weight, alcohol consumption, stress, and long-term physical inactivity (ILSI, 1990; Shils et al., 1994; Kannel, 1993). Coexisting conditions such as hypertension, elevated plasma cholesterol levels, and diabetes are to some extent preventable and reversible, and can be controlled by the individual through appropriate lifestyle modifications (Hoeger, 1987). Several dietary factors can have a significant impact on CAD risk factors that involve hyperlipidemia, hypertension, obesity, and glucose intolerance. The evidence is strong for the relationship between saturated fat intake, high blood cholesterol, and increased risk of CAD. High fat intake is also associated with increased risk of obesity. Dietary recommendations from various health agencies to reduce CAD risk are to reduce total fat and saturated fatty acids, cholesterol, and salt in the diet and to maintain a healthy body weight (USDHHS, 1988 and 1990).

Health risk appraisal describes a person's chances of becoming ill or dying from selected diseases. The model attempts to estimate mortality risk quantitatively (Goetz et al., 1980; Ellis and Raines, 1983). The average risk for all members in a group can be obtained from mortality tables, which express deaths as expected deaths over the next 10 years for a unit population of 100,000. Such tables have been developed by Harvey Geller and derived from U.S. National Center for Health Statistics mortality data. These mortality tables give the death rates for all causes combined and for each of the major causes by race, sex, and age group. The death rates are arranged in five-year age groups by race, sex, and age. The 1982 Geller tables from the Centers for Disease Control were used in this model (Hall and Zwemer, 1979; Center for Health Promotion and Education, 1982).

For most of the major causes of death there are well-recognized risk-related characteristics or precursors that have proved significant in assessing the prospective risks of death. These characteristics are called *risk indicators* or prognostic characteristics. A risk factor is the quantitative weight attached to a *risk indicator* (RI) to describe for a particular cause of disease or death the amount that the indicator increases or decreases the risk of death or disease (Goetz et al., 1980). The models use the results of studies that allow estimation of a *risk factor ratio* (RF). The RF for a particular risk indicator and event is the "odds ratio" for the event [p/(1 − p) where p is the probability of the occurrence of the event] in the members of a race-sex-age group with a particular risk indicator divided by the risk of the same event [p/1 − p)] in all members of the race-sex-age group:

$$RF = \frac{p/(1-p) \text{ in members of a race-sex-age group with a particular RI}}{p/(1-p) \text{ for all members of the race-sex-age group}}$$

The risk factors may also be obtained by dividing the mortality rate for the event among members of a race-sex-age group with a particular risk indicator by the mortality rate for the event in all members of the race-sex-age group. The risk factors for the average member of a race-sex-age group are always 1.0 (Hall and Zwemer, 1979; Chiang, 1984).

In some cases, p/(1 − p) values are obtained from logistic functions relating the natural logarithm (Ln) of the probability of the event to the associated risk indicator:

$$Ln \, (p/(1-p)) = a + bx_i \tag{15.3}$$

The odds ratio ($p/(1 - p)$) is determined from the appropriate values for the logistic intercept (a), the logistic regression coefficient (b), and the values of the risk indicator (x_i). Next, the age specific values for the Ln $p/(1 - p)$ at a particular level of the risk indicator are linearly regressed on age, so that:

$$\text{Ln } (p/(1 - p)) = A + B * \text{Age} \qquad (15.4)$$

where A and B are the linear intercept and linear regression coefficients respectively. The RF for the event at a designated age in years is calculated from the ratio:

$$RF = EXP (A + B * \text{Age})/ EXP (A' + B' * \text{Age}) \qquad (15.5)$$

The risk factor tables used in this model (Tables 15.1 and 15.2) are referred to as the Geller-Gesner tables (Hall and Zwemer, 1979; Raines and Ellis, 1982). A *composite risk factor* is an expression of risk in an individual for a particular cause of death, calculated from the individual risk factors applicable to that cause of death. The factors are combined as follows: All RF-values < 1 are multiplied together and this product is added to the sum of all risk factors > 1. The number of risk factors > 1 is subtracted from this sum. The resulting composite risk factor is multiplied by the population average incidence of the cause of death (Raines and Ellis, 1982). This gives an adjusted or individualized risk for the cause of death.

Survival advantage represents the reduction in an individual's present risks by intervention. It is the total percentage risk reduction and is calculated by dividing the amount of reduction by total individual risk (Hall and Zwemer, 1979).

TABLE 15.1. Male risk factor table

CAD RI	Age Groups										
	20	25	30	35	40	45	50	55	60	65	70
BP: Systolic											
200	5.3	5.3	5.3	4.8	4.3	3.9	3.6	3.2	2.9	2.7	2.4
180	3.2	3.2	3.2	3.0	2.7	2.5	2.3	2.1	2.0	1.8	1.7
160	2.0	2.0	2.0	1.9	1.7	1.6	1.5	1.4	1.3	1.2	1.1
140	1.2	1.2	1.2	1.2	1.1	1.0	1.0	0.9	0.9	0.8	0.8
120	0.8	0.8	0.8	0.7	0.7	0.7	0.6	0.6	0.6	0.6	0.5
BP: Diastolic											
105	1.5	1.5	1.5	1.6	1.6	1.7	1.8	1.8	1.9	1.9	2.0
100	1.3	1.3	1.3	1.4	1.4	1.4	1.5	1.5	1.6	1.6	1.7
95	1.1	1.1	1.1	1.2	1.2	1.2	1.3	1.3	1.3	1.3	1.4
90	1.0	1.0	1.0	1.0	1.0	1.0	1.1	1.1	1.1	1.1	1.2
85	0.8	0.8	0.8	0.9	0.9	0.9	0.9	0.9	0.9	0.9	1.0
80	0.7	0.7	0.7	0.7	0.7	0.8	0.8	0.8	0.8	0.8	0.8
Cholesterol											
280	2.2	2.2	2.0	1.8	1.7	1.5	1.4	1.2	1.1	1.0	1.0
220	0.6	0.6	0.6	0.7	0.7	0.8	0.8	0.9	0.9	1.0	1.0
180	0.2	0.2	0.3	0.3	0.4	0.5	0.6	0.7	0.8	1.0	1.0

TABLE 15.1. (*continued*)

CAD RI	Age Groups										
	20	25	30	35	40	45	50	55	60	65	70
Diabetic (Y)	7.4	7.4	6.9	6.3	5.8	5.4	5.0	4.6	4.2	3.9	3.6
Controlled	3.8	3.8	3.5	3.2	3.0	2.7	2.5	2.3	2.1	2.0	1.8
Not diabetic	1.0	1.0	1.0	1.0	1.0	1.0	1.0	1.0	1.0	1.0	1.0
Exercise											
Sedentary	1.0	1.0	1.1	1.2	1.2	1.3	1.3	1.4	1.5	1.6	1.6
Little	1.0	1.0	1.0	1.0	1.0	1.1	1.1	1.1	1.2	1.2	1.3
Moderate	1.0	1.0	1.0	0.9	0.9	0.9	0.9	0.9	0.9	0.9	0.9
Vigorous	1.0	1.0	0.9	0.9	0.9	0.8	0.8	0.7	0.7	0.7	0.7
Prescribe exercise for sedentary	1.0	1.0	1.0	1.0	1.0	1.0	1.0	1.0	1.0	1.0	1.0
Prescribe exercise for others	1.0	1.0	0.9	0.9	0.9	0.8	0.8	0.7	0.7	0.7	0.7
Current smoker 40+ cigarettes/day:	3.0	3.0	2.7	2.4	2.2	2.0	1.8	1.6	1.4	1.3	1.2
20-39	1.7	1.7	1.6	1.6	1.5	1.5	1.4	1.4	1.3	1.3	1.2
10-19	1.0	1.0	1.1	1.1	1.1	1.1	1.1	1.1	1.2	1.2	1.2
1-9	0.6	0.6	0.6	0.7	0.7	0.7	0.8	0.8	0.9	0.9	1.0
Former smoker 1-19 cigarettes/day:											
1 yr ago	0.7	0.7	0.7	0.7	0.7	0.8	0.8	0.9	0.9	1.0	1.0
1-4 yrs ago	0.5	0.5	0.5	0.5	0.5	0.6	0.6	0.7	0.7	0.8	0.9
10-17 yrs ago	0.4	0.4	0.4	0.4	0.4	0.4	0.5	0.6	0.6	0.7	0.8
Former smoker 20+ cigarettes/day:											
1 yr ago	1.0	1.0	1.0	1.0	1.0	0.9	0.9	0.9	0.9	0.9	1.0
1-4 yrs ago	0.9	0.9	0.9	0.9	0.9	0.9	0.9	0.9	0.9	0.9	0.9
10-19 yrs ago	0.6	0.6	0.6	0.6	0.6	0.7	0.7	0.7	0.8	0.8	0.9
20 yrs ago	0.4	0.4	0.4	0.4	0.4	0.4	0.5	0.6	0.6	0.7	0.8
Nonsmoker	0.3	0.3	0.3	0.4	0.4	0.4	0.5	0.6	0.6	0.7	0.8
Weight											
60% overweight	1.2	1.2	1.2	1.3	1.3	1.4	1.4	1.4	1.5	1.5	1.6
50% overweight	1.1	1.1	1.1	1.2	1.2	1.2	1.3	1.3	1.3	1.4	1.4
20% overweight	0.9	0.9	0.9	0.9	1.0	1.0	1.0	1.0	1.0	1.0	1.0
Average weight	0.8	0.8	0.8	0.8	0.8	0.8	0.8	0.8	0.8	0.8	0.8
10% underweight	0.8	0.8	0.8	0.8	0.7	0.7	0.7	0.7	0.7	0.7	0.7

TABLE 15.2. Female risk factor table

CAD RI	Age Groups										
	20	25	30	35	40	45	50	55	60	65	70
BP: Systolic											
200	5.3	5.3	5.3	4.8	4.3	3.9	3.6	3.2	2.9	2.7	2.4
180	3.2	3.2	3.2	3.0	2.7	2.5	2.3	2.1	2.0	1.8	1.7
160	2.0	2.0	2.0	1.9	1.7	1.6	1.5	1.4	1.3	1.2	1.1
140	1.2	1.2	1.2	1.2	1.1	1.0	1.0	0.9	0.9	0.8	0.8
120	0.8	0.8	0.8	0.7	0.7	0.7	0.6	0.6	0.6	0.6	0.5

TABLE 15.2. (continued)

CAD RI	20	25	30	35	40	45	50	55	60	65	70
						Age Groups					
BP: Diastolic											
105	1.5	1.5	1.5	1.6	1.6	1.7	1.8	1.8	1.9	1.9	2.0
100	1.3	1.3	1.3	1.4	1.4	1.4	1.5	1.5	1.6	1.6	1.7
95	1.1	1.1	1.1	1.2	1.2	1.2	1.3	1.3	1.3	1.3	1.4
90	1.0	1.0	1.0	1.0	1.0	1.0	1.1	1.1	1.1	1.1	1.2
85	0.8	0.8	0.8	0.9	0.9	0.9	0.9	0.9	0.9	0.9	1.0
80	0.7	0.7	0.7	0.7	0.7	0.8	0.8	0.8	0.8	0.8	0.8
Cholesterol											
280	1.2	1.2	1.2	1.2	1.1	1.1	1.1	1.1	1.1	1.1	1.1
220	0.8	0.8	0.8	0.8	0.7	0.7	0.7	0.6	0.6	0.6	0.6
180	0.7	0.7	0.6	0.6	0.5	0.5	0.5	0.4	0.4	0.4	0.4
Diabetic (Y)	39.6	39.6	32.4	26.6	21.8	17.8	14.6	12.0	9.8	8.0	6.6
Controlled	16.7	16.7	13.7	11.2	9.2	7.5	6.2	5.1	4.1	3.4	2.8
Not diabetic	1.0	1.0	1.0	1.0	1.0	1.0	1.0	0.9	0.9	0.9	0.9
Exercise											
Sedentary	1.0	1.0	1.1	1.2	1.2	1.3	1.4	1.5	1.6	1.7	1.8
Little	1.0	1.0	1.0	1.1	1.1	1.2	1.2	1.3	1.3	1.4	1.4
Moderate	1.0	1.0	1.0	0.9	0.9	0.9	0.9	0.9	0.9	0.9	0.9
Vigorous	1.0	1.0	1.0	1.0	1.0	0.9`	0.8	0.8	0.7	0.7	0.6
Prescribe exercise for sedentary	1.0	1.0	1.0	1.0	1.0	1.0	1.0	1.0	1.0	1.0	1.0
Prescribe exercise for others	1.0	1.0	1.0	1.0	0.9	0.9	0.8	0.8	0.7	0.7	0.6
Current smoker											
20–39 cigarettes/day	3.2	3.2	2.9	2.7	2.4	2.2	2.0	1.8	1.7	1.5	1.4
10–19	1.9	1.9	1.8	1.7	1.6	1.5	1.4	1.4	1.3	1.2	1.2
1–9	0.9	0.9	0.9	0.9	0.8	0.8	0.8	0.8	0.8	0.8	0.8
Nonsmoker	0.5	0.5	0.5	0.6	0.6	0.6	0.7	0.8	0.8	0.8	0.8
Weight											
60% overweight	5.9	5.9	4.6	3.5	2.7	2.1	1.6	1.2	1.0	1.0	1.0
50% overweight	3.9	3.9	3.2	2.6	2.2	1.8	1.4	1.2	1.0	1.0	1.0
20% overweight	1.1	1.1	1.1	1.1	1.1	1.1	1.0	1.0	1.0	1.0	1.0
Average weight	0.5	0.5	0.5	0.6	0.7	0.7	0.8	0.9	1.0	1.0	1.0
10% underweight	0.3	0.3	0.4	0.4	0.5	0.6	0.7	0.9	1.0	1.0	1.0

15.5 STELLA® Model for CAD Risk

The model shown in Figure 15.3 was developed using **STELLA®** software. The population stock was set at an arbitrary value of 10,000,000. CAD deaths, the outflow from the population stock, represents the number of deaths from CAD calculated as the product of the population and the CAD death rate. The converter, CAD death rate, describes an individual's chances of dying from CAD. It varies depending on the values of the two user inputs, gender and composite risk factor.

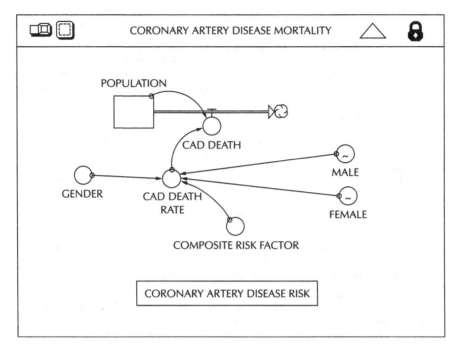

CORONARY ARTERY DISEASE MORTALITY

POPULATION

CAD DEATH

GENDER   CAD DEATH
                  RATE

MALE

FEMALE

COMPOSITE RISK FACTOR

CORONARY ARTERY DISEASE RISK

FIGURE 15.3

In the gender input icon, the user enters either 1 for male or 2 for female. Logical functions assign the appropriate male or female CAD mortality rate based on data stored in the male and female convertor icons. The CAD mortality data were obtained from the Geller tables (Center for Health Promotion and Education, 1982). In this model, the CAD death rates for white males and white females are entered as deaths per 100 over the next 10 years and stored as a graphical function according to time expressed as age in years. The resulting CAD death rate equals the product of this gender-specific CAD death rate and the composite risk factor.

The composite risk factor for the average population is equal to 1.0. The user can enter a new composite risk factor value calculated from the CAD risk factors in Figure 15.4. The individual risk-factor values are obtained from the appropriate male or female risk factors derived from the Geller-Gesner tables. These tables list the lifestyle-related risk factors for the different risk indicators according to 5-year age groups. To simplify the present model, the detailed tables are not included; rather, general values will be used.

The time specifications for simulation are set between the ages of 20 and 75 years. The output displays the estimated probabilities of dying from CAD in the next 10 years for a population unit of 100. The comparative selection allows the user to simultaneously view and compare the average population CAD risk, the CAD risk with several risk factors present, and risk reduction through lifestyle modification of these risk factors. When the lesson is completed, the dynamite tool is used to clear both of these pads.

FIGURE 15.4. Instructions for coronary artery disease worksheet

Present State:
  I   Enter values of **Present Risk Indicators** in the first column.
  II Enter **Risk Factors** for the risk indicators using **Table 15.1 or 15.2**. When a risk indicator falls between two given risk factors, it may be necessary to interpolate or estimate. For **Blood Pressure**, use the higher risk factor of systolic and diastolic. However, if both are over 1.0, then use both risk factors.
  III Calculate the **Composite Risk Factor** as follows:
     1. Record the **Risk Factors** in the two columns, X and +.
     2. In the left-hand column (X) record all the **Risk Factors** that are **1.0 or less**. Also, record the average 1.0 for all other risk factors.
     3. In the right-hand column (+) record all the **Risk Factors greater than 1.0** but **subtract 1.0** from each risk factor before recording it. (The 1.0 is already in the left-hand column).

       Example: For a risk factor of 2.7, record   X   +
                                     1.0   1.7   (2.7−1.0=1.7)

     4. **Multiply** the factors in the left-hand column (X) by each other. Enter the product.
     5. **Sum** the factors in the right-hand column (+). Enter the sum.
     6. **Add** the **product (X) to the sum (+)** to obtain the composite risk factor.

Intervention/Risk Reduction:
  I   Identify and enter achievable alterations in the risk indicators that will favor a reduction in CAD risk into the **Reduced CAD RI** column. Carry over unchanged risk indicators.
  II Enter the **Risk Factors** associated with these risk indicators as described above.

     Blood Pressure: Base the risk factor after intervention on the average of the previous high reading with the anticipated reading after intervention.

  III Calculate the **Composite Risk Factor** following the same instructions as above (III).

To operate the model, the user enters either 1 for male or 2 for female in the gender icon, runs the model, and views the average population CAD risk or mortality using the graph pad or the table pad. To examine the effects of lifestyle on CAD risk, the user enters the values of the risk indicators and the corresponding risk factors.

Finally, *risk reduction* from lifestyle modification can be simulated. This is done by improving certain lifestyle-related risk indicators and repeating the procedures in the second step. Risk reduction illustrates how positive behaviors potentially affect one's lifespan. To conclude the lesson, the user clears the graph and table pads with the dynamite tool, and sets the composite risk factor back to 1.

15.6 Simulation Results

The gender input and composite risk factor input have the most control over the system dynamics. First, test the effects of gender on CAD risk. Set the composite risk factor to 1 representing the average population when no risk factors are considered, and enter either 1 or 2 to indicate gender. The simulation shows that males have a higher probability of dying from CAD than fe-

FIGURE 15.4. (*continued*) Coronary artery disease risk worksheet

Name _____

Age _____   Gender _____

		Present state				Intervention		
	Present		Composite		Reduced		Composite	
CAD Risk	CAD	Risk	Risk Factor		CAD	Risk	Risk Factor	
Indicators (RI)	RI	Factors	X	+	RI	Factors	X	+
Blood pressure								
Systolic								
Diastolic								
Cholesterol								
Diabetes								
Controlled?								
Exercise*								
Smoking								
Amount/day								
If stop, no. years ago								
Weight (%)**								
Composite Risk Factor	(X) + (+)				(X) + (+)			

*Daily Exercise: Sedentary—Walking less than 5 blocks.
                   Little—Walking 5-15 blocks.
                   Moderate—Walking 15-20 blocks.
                   Vigorous—Walking more than 20 blocks.

**Weight (%):   Use the weight Tables 1A, 1B, and 1C to determine the percent over- or underweight.

males. Also, CAD risk increases with age for both genders. Operation is simple; enter 1 or 2 in the Gender input device on the high level map shown in Figure 15.5, and enter the value for the Composite Risk Factor. The graph is set to operate in comparative mode, so the results of each simulation will be saved until the graph pad is cleared.

Next, the effects of risk factors on CAD death rate may be tested. Total blood cholesterol levels of 180, 220, and 280 mg per dL correspond to composite risk factors of 0.4, 0.7, and 1.7, respectively. As blood cholesterol increases, an individual's CAD risk increases. This evidence suggests that interventions to decrease blood cholesterol levels will reduce CAD risk.

Other individual risk factors include smoking 20 cigarettes per day, a total blood cholesterol level of 280 mg per dL, and a blood pressure of 180/90 mm

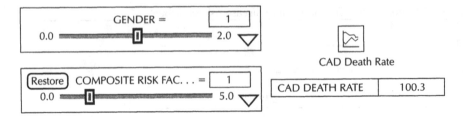

FIGURE 15.5

Hg. Individually, these correspond to composite risk factors of 1.5, 1.7, and 2.7. The results illustrate that different risk factors affect CAD risk to different degrees. The numerical values show that a blood pressure of 180/90 mm Hg increases CAD risk the most, followed by the elevated blood cholesterol, and then cigarette smoking (20/day). By entering these values in the CAD model, the user will find that the outcome depends on the multiple of risk indicated by the composite risk factor.

Interactions among individual risk factors cause a much higher degree of risk. Consider the following cases: (a) cigarette smoking (20/day); (b) cigarette smoking (20/day) and total blood cholesterol of 280 mg per dL; and (c) cigarette smoking, blood cholesterol of 280 mg per dL, and a blood pressure of 180/90 mm Hg. These cases correspond to composite risk factors of 1.5, 2.2, and 3.9, respectively. The outcomes in the model show that CAD risk increases as the number of risk factors increases; the effect is not simply additive but rather is compounded (Fig. 15.6). Therefore, an individual who reduces only one of several risk factors can significantly reduce CAD risk.

Now consider the concept of risk reduction. A 40-year-old male who smokes 20 cigarettes per day, has a blood cholesterol of 280 mg/dL, and a blood pressure of 180/90 mm Hg has a composite risk factor of 3.9. Achievable alterations in the risk indicator that favor risk reduction are: to stop smoking, to lower cholesterol to 220 mg/dL, and to lower blood pressure to 140/90 mm Hg. The new composite risk factor is 0.8. Compare the outcomes predicted by the CAD model with these values; the data clearly show the great impact of current lifestyle on chances of dying from CAD, as well as the improvement due to risk reduction. The difference represents *survival advantage*. Recall that the model predicts survivorship for a population, and that it is only possible to say that an individual can decrease risk, not that a particular person will survive.

Last, consider a 40-year-old female who smokes 10 cigarettes per day, has a blood cholesterol of 280 mg/dL, a blood pressure of 180/90 mm Hg, and is 20% overweight. Her individualized composite risk factor is 3.5. Achievable alterations in her risk indicator that favor risk reduction are: stopping smoking, lowering blood cholesterol to 220 mg/dL, lowering blood pressure to 140/90 mm Hg, and reaching an average weight. The new composite risk factor is 0.59. A simulation clearly demonstrates the impact of current lifestyle on chances of dying from CAD, plus the benefits of risk reduction.

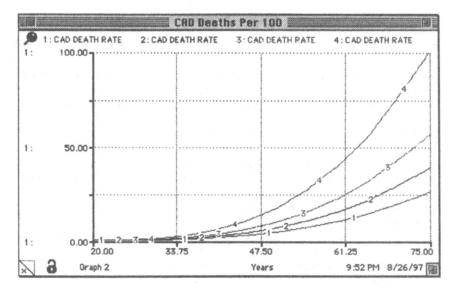

Figure 15.6

15.7 Discussion of the Model

This model simulates an individual's CAD health risk and quantitatively describes the relationship of an individual's gender and lifestyle on life span. The simulation results with the composite risk factor set to 1.0 correspond to published mortality data (Center for Health Promotion and Education, 1982), and the model is intended to convey the message that risk reduction through intervention can significantly affect chances for survival (Neaton and Wentworth, 1992).

Any attempt at health promotion is an attempt to alter behavior. In the absence of a means to employ the published information, the idea of health risk is qualitative. Therefore, this project developed a computer-assisted, simulation method designed to show the quantitative relationship of lifestyle behaviors on risk for CAD. As an educational tool, it is intended to aid in understanding how lifestyles affect health and the importance of modifying the undesirable aspects of behavior. The model could be useful in health education aiming at the modification of unhealthy behaviors and in motivating change (Farquhar et al., 1990).

References

Brown, M.L., ed. *Present Knowledge in Nutrition*, 6th ed. International Life Sciences Institute, Nutrition Foundation, 1990, 349–354.

Center for Health Promotion and Education, Centers for Disease Control, DHHS. *Probability of Dying Within the Next 10 Years at Selected Ages for 40 Major Causes*. Atlanta, Ga., April 1982.

Chiang, C.L. *The Life Table and Its Applications*. Malabar: FL, Robert E. Krieger, 1984; 16-17.

Ellis, L.B., and J.R. Raines. "Health Risk Appraisal: A Tool for Health Education." *Health Education* (Oct 1983), 30-34.

Farquhar, J.W., S.P. Fortmann, J.A. Flora, et al. "Effects of Communitywide Education on Cardiovascular Disease Risk Factors." *JAMA*, 264 (1990): 359-365.

Goetz, A.A., J.F. Duff, and J.E. Bernstein. "Health Risk Appraisal: The Estimation of Risk." *Health Promotion at the Worksite*, 95 (1980): 119-126.

Hall, J.H., J.D. Zwemer. *Prospective Medicine*. Indianapolis: Methodist Hospital of Indiana, 1979.

Hoeger, W.W.K. *The Complete Guide for the Development and Implementation of Health Promotion Programs*. Englewood, Colorado: Morton Publishing Co., 1987, 1-5, 51-69.

Kannel, W.B. "CHD Risk Factors: A Framingham Study Update." *Hospital Practice*, (July 1990), 119-130.

Kannel, W.B. "Long-Term Epidemiologic Prediction of Coronary Disease, The Framingham Experience." *Cardiology*, 82 (1993): 137-152.

MRFIT Research Group. "Multiple Risk Factor Intervention Trial, Risk Factor Changes and Mortality Results." *JAMA*, 248 (1982): 1465-1477.

MRFIT Research Group. "Mortality Rates After 10.5 Years for Participants in the Multiple Risk Factor Intervention Trial." *JAMA*, 263 (1990): 1795-1800.

National Research Council. *Diet and Health, Implications For Reducing Chronic Disease Risk*. Washington, D.C.: National Academy Press, 1989, 5-8, 99-103, 698.

Neaton, J.D., and D. Wentworth. "Serum Cholesterol, Blood Pressure, Cigarette Smoking, and Death From Coronary Heart Disease." *Arch. Intern. Med.*, 152 (1992): 56-63.

Raines, J.R., and L.B.M. Ellis. "A Conversational Microcomputer-Based Health Risk Appraisal." *Comp. Progr. Biomed.*, 14 (1982): 175-184.

Riggs, J.E. "The dynamics of aging and mortality in the United States, 1900-1988." *Mech. Aging Devel.* 66 (1992): 45-57.

Shils, M.E., J.A. Olson, and M. Shike. *Modern Nutrition in Health and Disease*, 8th ed. Philadelphia: Lea & Febiger, 1994, 1533-1544.

Strehler, B.L. *Time, Cells, and Aging*, 2nd ed. New York: Academic Press, 1977.

Strehler, B.L., and A.S. Mildvan. "General theory of mortality and aging." *Science* 132 (1960): 14-21.

Sytkowski, P.A., W.B. Kannel, and R.B. D'Agostino. "Changes in Risk Factors and the Decline in Mortality from Cardiovascular Disease, The Framingham Heart Study." *N. Engl. J. Med.*, 322 (1990): 1635-1641.

United States Department of Health and Human Services, Public Health Service. *The Surgeon General's Report on Nutrition and Health*. Washington, D.C.: U.S. Government Printing Office, 1988, 2-10, 22, 83-87.

United States Department of Health and Human Services, Public Health Service. *Healthy People 2000, National Health Promotion and Disease Prevention Objectives*. Washington, D.C.: U.S. Government Printing Office, 1990, 3, 19-21, 112-113.

16

Kinetic Genetics: Compartmental Models of Gene Expression

> Decay of RNA must have evolved in response to the evolution of the RNA synthetic capacity, *otherwise the world would have been clogged with macromolecules* and no new cells or any other biological entities could have developed.
>
> David Apirion et al.,
> "RNA Processing and Degradation,"
> *Biochem. Soc. Trans.* 14 (1986): 807–810

In Chapter 7, the topic of biodynamics was introduced in reference to changes in the rate of messenger RNA production using a single-compartment model. If the point of gene expression is to convert the latent information in DNA into the protein catalysts that conduct most of the business of the cell, then it is necessary to link any change in messenger RNA production to a change in protein synthesis. This chapter demonstrates a model of gene expression that includes production of mRNA and synthesis of protein. Chapter 17 extends these ideas to the level of the phenotype and develops a more comprehensive model of gene expression.

16.1 Using the Idea of Approximation to Simulate Gene Expression

Kinetic models of gene expression have always been expressed as analytical solutions to differential equations. An alternative is to use **STELLA®** to create corresponding finite difference equations, and to solve them using short calculation intervals (dt or DT in the Time Specifications under the Run menu). Let us first compare the two methods and determine whether similar results can be obtained. The simplest kinetic model of gene expression based on the assumptions of zero-order synthesis and first-order decay (Berlin and Schimke, 1965; Schimke, 1973; Yagil, 1975) can be represented as a pool of translationally active mRNA that encodes a specific protein (Fig. 16.1A). In this model, the rate of mRNA synthesis (λ_1) is equivalent to the rate at which active mRNA appears in the cytoplasm, and represents an aggregate rate of transcription, processing, and nucleocytoplasmic transport. The rate of protein synthesis (λ_3) is assumed to be linearly proportional to mRNA concentration, and the protein and mRNA are degraded with first-order kinetics (λ_2 and λ_4, respectively). From these assumptions, the following set of equations

A. Two-Compartment Model of Gene Expression

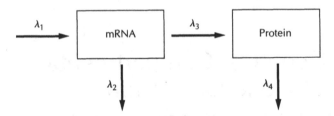

B. Two-compartment model represented in STELLA®

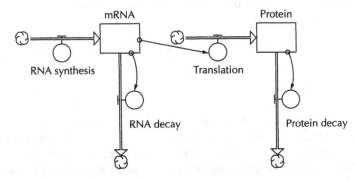

FIGURE 16.1. (A) Traditional depiction of a two-compartment model of gene expression, with rates of synthesis and decay indicated by arrows flowing into and out of the compartments. (B) The same model depicted with the user interface for the **STELLA®** simulation program. The arrows shown in the upper diagram are replaced by icons called flows, and proportionality, such as first-order decay, is represented by arrows called connectors. Each circle and rectangle is associated with a dialog box for entering numbers or equations.

may be derived, in which R indicates the concentration of a specific mRNA and P designates the concentration of the protein encoded by that mRNA. R_0 is defined as the concentration of mRNA in the system during the initial steady-state, R_t the amount at any moment between initial and final steady states, and R_{ss} the concentration at the new steady state; protein concentrations at different times are designated by the same subscripts. The rate parameters for synthesis and degradation are indicated by the letter λ, with subscripts as shown in Figure 16.1:

Equation 16.1: The rate of change in concentration of a gene product between two steady-state levels is a function of its rates of synthesis and decay, assuming zero-order synthesis and first-order decay:

$$\frac{dR}{dt} = \lambda_1 - \lambda_2 R_t \qquad (16.1)$$

Equation 16.2: The solution to Eq. 16.1 shows that the time course required for the concentration of a gene product to attain a new steady state is related to the decay constant or half-life, whether the concentration is increasing or decreasing:

$$R_t = R_0 + (R_{ss} - R_0)(1 - e^{-\lambda 2^t})$$ (16.2)

Equation 16.3: In the initial and final steady states, the rate of synthesis equals the rate of decay (which is the product of the rate constant for decay and the concentration of product). The initial and final concentrations of mRNA equal the ratio between the two values in effect during the initial and final steady states:

$$\lambda_1 = \lambda_2 R_{ss} \quad \text{or} \quad R_{ss} = \frac{\lambda_1}{\lambda_2}$$ (16.3)

An equation may be written to show the relationship of protein concentration to mRNA concentration using the assumptions that protein synthesis is related to mRNA concentration by a first-order rate constant for translation and that protein decay follows first-order kinetics. In this solution, the rate constants for synthesis and decay of mRNA are designated λ_1 and λ_2, respectively, and the corresponding values for protein are designated λ_3 and λ_4.

Equation 16.4: The concentration of a protein may be expressed in terms of the concentration of the corresponding mRNA:

$$P_t = P_0 + (P_{ss} - P_0)(1 - e^{-\lambda 4^t})$$
$$- \lambda_3(R_{ss} - R_0)\left\{\frac{e^{-\lambda 4^t} - e^{-\lambda 2^t}}{\lambda_2 - \lambda_4}\right\}$$ (16.4)

Note that the first two terms on the right side of Eq. 16.4 are identical in form to the solution to the single-compartment model (Eq. 16.2). The last term on the right introduces a delay between induction of mRNA and induction of the corresponding protein.

A traditional computer program based on the two-compartment model of gene expression solves Eq. 16.2 and 16.4 when initial values for the parameters are supplied, along with one or more changes from the original steady state (Hargrove and Schmidt, 1989). The parameters needed are readily calculated from experimentally determined concentrations and half-lives, as shown in Table 16.1. However, it is worth noting that any implications of Eq. 16.4 would be lost on anyone who had no means of demonstrating the relationships it predicts. Worse yet, as more compartments were added to include items such as translational control of protein synthesis or nuclear mRNA dynamics, the equations would rapidly become very complex and the time required to carry out the work would grow. Now compare this method with the mode of operation and results obtained using **STELLA®** for computer simulation using compartmental analysis.

Before attempting this, permit me to make one more observation. Equation 16.4 was originally derived by Dr. David Hoel (see Appendix to Lee et al.,

TABLE 16.1. Example of calculations for kinetic parameters and steady-state values when modeling gene expression.

	Rate of synthesis (units/h)	Half-time of decay (h)	Rate constant (h^{-1})	Half-time of transit (h)	Rate constant (h^{-1})	Concentration
Primary transcript	1.0[a]	1.0	0.693[b]	0.16	4.15	0.21
Nuclear mRNA	0.87[c]	0.33	2.08	0.33	2.08	0.21
Cytoplasmic mRNA	0.44	2.0	0.347	N.A.[d]	N.A.	1.26
Protein	126[e]	4.0	0.173	N.A.	N.A.	728

[a] The rate of synthesis in this example was assumed to be 1 unit per hour; units could be molecules • $(cell^{-1} \cdot h^{-1})$ or moles • $(grams of tissue^{-1} \cdot h^{-1})$.
[b] The kinetic parameter equals 0.693/half-life.
[c] The rate of synthesis at steady state for each intermediate equals the rate constant for conversion of the preceding constituent times the concentration of that constituent.
[d] N.A., not applicable.
[e] In this example, the rate of protein synthesis was arbitrarily set at 100 moles of protein per mole of mRNA per hour, which is low compared to rates in tissues such as liver.

1970) and published in the *Journal of Biological Chemistry*. It would be impossible for anyone to derive Eq. 16.4 who had not completed at least one course in differential equations (or more likely, who had earned a doctorate in Biometrics), and this result is only one possible solution that holds under one set of strict assumptions that are never attained in a living animal. In the next section, we will set up a model that could be created by any student who had completed high school algebra, and the outcome will be equally valid.

16.2 Solving a Model of Gene Expression with a Simulation Program

In order to compare results of the traditional approach with those obtained by the method of approximation, the block diagram of gene expression represented in Figure 16.1A was converted to a computer program using **STELLA®** (Hargrove et al., 1993). The equations were not entered just by copying the solutions previously listed, because the stock-and-flow diagrams produce asymptotic growth properly when the algebra is set up correctly in the dialog boxes. For example, the initial RNA concentration at steady state equals the rate of synthesis divided by the rate constant for decay, and this value may be calculated and typed into the highlighted line. Because the traditional model assumes an instantaneous increase in RNA synthesis, the STEP function from the Built-ins list was used to provide this feature.

The model of gene expression shown in Figure 16.1 is based on the assumptions that the rate of mRNA synthesis is zero-order (i.e., equal amounts are produced in each interval), whereas protein synthesis and decay processes are first-order (i.e., proportional to concentration of mRNA or protein). To indicate that flows are proportional to existing mRNA or protein us-

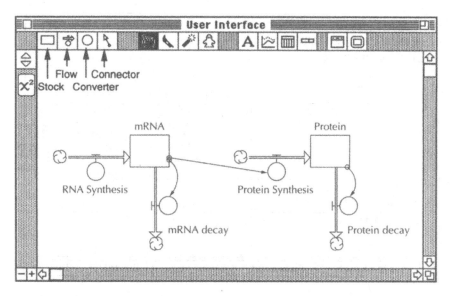

FIGURE 16.2. A two-compartment model of gene expression was created using **STELLA®**. Note that the rates of degradation of the mRNA and the protein are proportional to concentration (first order) as indicated by the connectors linking the stocks to the outflows.

ing **STELLA®**, a connector is used to join the stock to the valve on the outflow (Fig. 16.2). Because most proteins and mRNAs decay in this fashion, *the time scale of accumulation for mRNAs and proteins is thought to be determined largely by their rates of decay*. According to the kinetic model of Figure 16.1, the major reason for the delay between initiation of transcription and accumulation of protein is that a new steady state in product concentration cannot be attained until five half-lives have elapsed. To duplicate this feature, decay constants were defined as follows:

$$\lambda = \ln 2/\text{half-life} = 0.693/\text{half-life} \qquad (16.5)$$

For example, a half-life of 1 hour is equivalent to a decay constant of 0.693 h^{-1}. As an aid to setting the scale in graphs generated with **STELLA®**, it is useful to calculate the initial and final steady-state values of mRNA (R_{ss} and protein (P_{ss}). Notice from Figure 16.1 that the quantity of material in any compartment at steady state equals the rate of inflow divided by the sum of all applicable rate parameters for outflow. In the model shown in Figure 16.1, each compartment has only one outflow, and Eq. 16.3 gives the steady-state concentration of mRNA. The corresponding relationship for the protein is:

$$P_{ss} = R_{ss} \, (\lambda_3/\lambda_4) \qquad (16.6)$$

where λ_3 is the amount of protein synthesized per molecule of mRNA per unit of time and cell mass, and λ_4 is the rate of protein degradation calculated in the same manner as shown in Eq. 16.5.

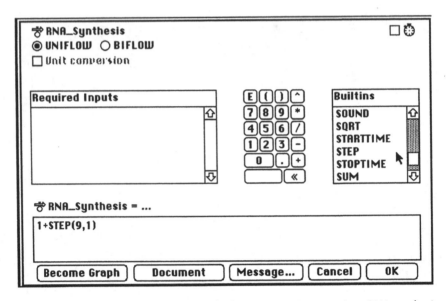

FIGURE 16.3. A dialog box is shown in which a stepwise increase in mRNA synthesis has been programmed to simulate mRNA induction.

To simulate gene expression, the kinetic model shown in Figure 16.2 only requires that four kinetic parameters be known, including the rate of mRNA synthesis, the mRNA's half-life and rate of translation, and the protein's half-life. In order to verify that **STELLA®** would produce the same results as our prior program, values were compared for the same kinetic constants and calculation intervals (0.25 h and 0.05 h). A change in the rate of synthesis was introduced by adding a STEP function that increased the rate of mRNA production in the dialog box (Fig. 16.3). When the shorter calculation interval was used, the maximum difference among values compared to results from the analytical solution equalled 0.05, or five parts per thousand for the hypothetical mRNA (Fig. 16.4A) and the protein (Fig. 16.4B). It is essential that users of this program test the effects of changing the calculation intervals to ensure that minimum values have been achieved. It is worthwhile noting that biological scientists are not capable of measuring differences with this degree of precision; measurement errors are more typically on the order of 5–10%! Now let us consider some of the implications of this simple model.

16.3 The Necessary Balance of Synthesis and Degradation

In adult mammals, the rates of total mRNA synthesis and protein synthesis are precisely balanced with the rates of mRNA degradation and protein degradation. For instance, the body weight of adult humans typically varies by no more than 1 pound per year between the ages of 20 and 50. Although this

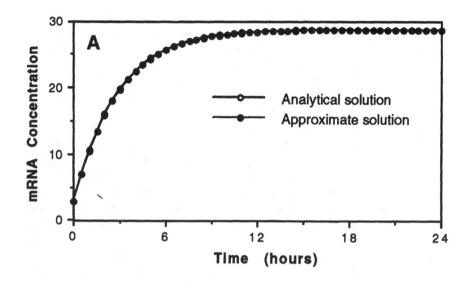

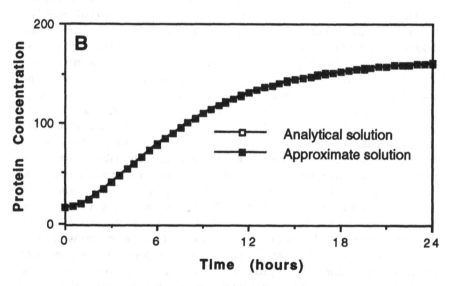

FIGURE 16.4. Results of simulations are compared for a program based on an analytical solution to the two-compartment model and the approximate solution obtained with the **STELLA®** package. The largest difference observed when the calculation interval equalled 0.025 h was 0.5% of the analytical solution, which cannot be detected when the results are plotted together.

observation may seem commonplace, it is very difficult to achieve because the processes involved in the biosynthesis of macromolecules are chemically and mechanistically distinct from the process of degradation. In other words, the body weight of a healthy, adult human is very nearly in a steady state, and extensive physiological mechanisms have evolved to support this balance. Although considerations of mass balance are very simple, they disguise an underlying complexity. The cells that make up each organ individually synthesize from 3,000 to 12,000 different gene products, and the processes involved in mRNA synthesis and protein synthesis are elegantly complex. Knowing this, is it likely that the degradative processes are any less elegant? The balance equations indicate that the total amount of mRNA transcription, total production of mature, cytoplasmic mRNA, and the synthesis of all the proteins needed to enable each organ to carry out its vast repertoire of functions are precisely balanced by the total amount of degradation. On a typical day, the balance is exact to within a fraction of a percent, and if the individual is healthy and well nourished, the balance continues for years. If this were not so, the individual would continue to grow, often with deleterious results if nutritional requirements exceeded availability. At the cellular level, the choice for growth is between hypertrophy and cell division, and each process eventually becomes disadvantageous to a mammal, if not catastrophic.

What is the simplest way to effectively bring the system into balance? The answer is to achieve a rate of product degradation that is proportional to the amount of product present—in other words, first-order elimination! This automatically puts limits on growth, and can be achieved in one of three ways: either the cell must divide, the mRNA and proteins must be secreted, or the gene products must be eliminated by chemical degradation and recycling. Bacteria are unicellular organisms, and the content of most proteins is balanced by a combination of rapid mRNA decay and frequent cell division. Adult mammals, which do not have the luxury of unlimited growth, utilize rates of mRNA and protein degradation that effectively balance synthesis. From this perspective, proportional rates of product elimination had to be invented, and the task may have been just as difficult as developing precise mechanisms of cell cycle control. Despite the well-known difficulty in obtaining stable preparations of most mRNAs and some proteins, the molecules are extremely stable in the absence of degradative enzymes when protected from oxidative destruction. These considerations lead to a series of predictions about mRNA and protein metabolism that are based on the essentials of cellular and organismal homeostasis, plus chemical balance and optimal cell size.

16.4 Transcriptional Controls Are Most Efficient

Mathematically, controls over synthesis or degradation can produce equivalent changes in protein concentration, but transcriptional controls are most efficient due to the small scale involved. Each autosomal gene is present at a very low concentration (two copies per cell), and the use of sequence-specific DNA-binding proteins permits one variety of protein to control many genes. The amplification involved in gene expression is enormous, because genes are present at a concentration of not over 10^{-12} moles per kg of tissue. This estimate was obtained by dividing the number of hepatocytes in liver (about 10^{11} hepatocytes per kg) by Avogadro's number. Whereas each of the two alleles at a genetic locus can be regulated by a single molecule of a DNA-binding protein, the corresponding mRNA may be represented by hundreds or thousands of copies, and therefore has a concentration in the range of 10^{-10} moles/kg. A factor that bound mRNA stoichiometrically would need to be present at a much higher concentration than a DNA-binding protein to achieve the same effect on protein synthesis. Because of the amplification that takes place during protein synthesis, it is plainly not feasible to control protein turnover by stoichiometric binding to a second protein except in special cases. Liver contains about 0.16 g of protein per gram tissue; assuming an average molecular weight of 40,000 grams per mole, the concentration of an average protein equals 4×10^{-7} moles per kg. The relative abundance of proteins compared to genes in liver is therefore about 10^6 moles of protein per mole of gene. Enzymes control the concentration of metabolites, which may range from about 10^{-6} to 10^{-3} moles/kg. The relative excess of metabolites relative to enzymes makes it feasible to control proteins by ligand binding or covalent modification. Both methods are used to regulate cholesterol synthesis by hydroxymethylglutaryl (HMG) CoA reductase, which is unstable in the presence of sterols (Goldstein and Brown, 1990), and is inactivated by an AMP-dependent protein kinase (Parker et al., 1989). From the perspective of scale, an enzyme is a gene's means of controlling utilization of metabolites that are present in vastly greater concentration, as appropriate to energy metabolism.

There is a drawback to controls that alter transcription alone. Because mammalian cells do not divide frequently, and because most mRNAs and proteins are relatively stable, the preformed molecules persist in tissues for several hours or days after synthesis is terminated. When it is essential that timing be precisely controlled, covalent modification or targeted degradation must be employed to remove the preexisting products. Bacteria have other options because rapid cell division dilutes preexisting proteins, and it is evident that mammals employ many more kinds of posttranscriptional regulation than do bacteria. An excellent example is cyclic AMP-dependent protein kinase, which controls glucose metabolism in mammals by regulating glycogen synthase and glycogen phosphorylase (Pilkis et al., 1988). Instead of using a protein kinase for this regulation, most microbes employ a cyclic AMP-binding protein called catabolite activator protein, which controls glucose utilization through effects on transcription (DeCrombrugghe et al., 1984). Kinetic considerations suggest that mammals invented protein kinases and phosphatases in order to solve the problem of slow adaptation that is imposed by a long life span. Glycogen must be mobilized in seconds during an emergency, and this cannot be achieved by synthesizing an enzyme with a half-life of many hours. A fundamental principle of systems theory is that delays between a signal and a response beget instability.

16.5 Rapidly Inducible Proteins are Encoded by Labile mRNAs

Because the approach to steady state depends on a gene product's half-life, a delay will always occur between the time an mRNA attains its half-maximal response and the time when the encoded protein attains its half-maximum value. The mRNA concentration is usually a good index of protein synthesis, but it is akin to a faucet that is filling the protein pool, and not the amount of material already in the pool. Plainly, much more time is required to fill a large pool (stable protein) than a small one (unstable protein). Rapid degradation of mRNAs and proteins is an adaptive feature that permits rapid regulation of concentration, but to say this begs the question of how relative stability evolved. To the degree that stability depends on physical structures and enzyme-based mechanisms such as ubiquitin-protein degradation, then these features must have evolved by natural selection to facilitate interaction of specific gene products with degradative systems.

One example of differences in structural features that are related to stability is shown in the comparison of tyrosine aminotransferase and aspartic aminotransferase in Figure 16.5. Aspartic aminotransferase is a very stable en-

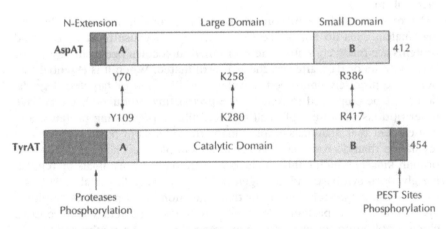

FIGURE 16.5. The structure of the stable enzyne, aspartic aminotransferase, is compared with the structure of its unstable cousin, tyrosine aminotransferase. Both enzymes contain specific amino acid residues at conserved positions; these serve the catalytic functions of the two enzymes. Tyrosine aminotransferase differs in having segments at each end that are sensitive to degradative and regulatory enzymes.

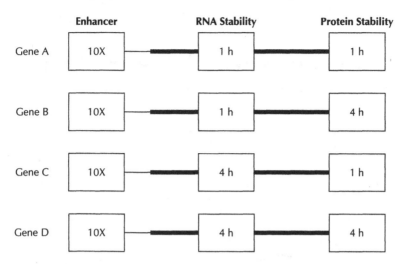

FIGURE 16.6. The interval required for an increased level of mRNA to cause a proportional increase in concentration of the encoded protein is predicted to be related to the protein's half-life. This can be demonstrated using **STELLA®** by comparing four genes A–D, whose transcription is regulated to the same degree by binding of regulatory factors to enhancer elements (left side of figure). The genes differ as shown by the stability loci indicated for each RNA and protein.

zyme, whereas tyrosine aminotransferase has a half-life of about 2 h. The enzymes are similar in the positions of a number of amino acid residues that are involved in catalytic functions, indicated as Y70/109, K258/280, and R386/417 (these are tyrosine, lysine, and arginine, respectively). However, it has now been demonstrated that tyrosine aminotransferase possesses sites at each end of the molecule that cause it to be sensitive to proteolytic degradation (probably by a ubiquitin-dependent protease that is part of a degradative structure called the proteasome).

In order to coordinate induction of mRNAs and proteins, the half-lives of each kind of molecule must be short. An example of the principle is shown in Figure 16.6. Imagine four different genes whose transcription can be increased tenfold by a common transcription factor, but that produce mRNAs and proteins with the half-lives indicated. Figure 16.7A shows the very short predicted delay between an increase in the mRNA with a 1-h half-life and the increase in the concentration of an unstable protein. Compared to this, the accumulation of the more stable protein in Figure 16.7B is delayed and only attains its final concentration much later due to its 4-h half-life. If the mRNA has a longer half-life, its induction is also delayed (Figure 16.7C), and if the RNA and protein are both stable, the full increase will not occur for more than a day because five half-lives of the RNA must elapse, followed by five half-lives of the protein (Figure 16.7D).

The idea that decay constants control the time course of induction to new steady-state levels applies independently to mRNAs and proteins because

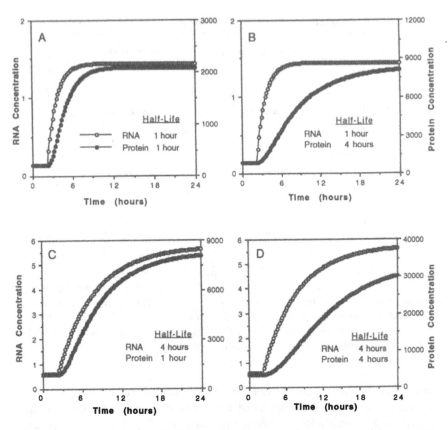

FIGURE 16.7. Outcomes of STELLA® simulations for the four genes shown in Figure 16.6 are plotted here. (A) Rapid induction and minimal delay between increase in mRNA and protein are predicted for gene A. (B) A longer delay is predicted for the protein with the longer half-life that is encoded by gene B. (C) An RNA with a longer half-life will be induced later than the less stable product of gene A, but will achieve a higher concentration. (D) If the mRNA and protein are both stable, induction is slow and a delay occurs between the increase in mRNA and protein. However, the protein achieves a much higher relative concentration compared to the other cases.

these molecules represent separate pools. Because most proteins are more stable than the corresponding mRNAs, the time required for proteins to increase to new levels will be delayed in comparison to their mRNAs by a value related to the half-life of the protein. This is confirmed by experimental results. The mRNA for tyrosine aminotransferase begins to increase within a few minutes after treatment of hepatoma cells with dexamethasone phosphate (a synthetic, glucocorticoid hormone), and within two hours this gives rise to an increased concentration of the enzyme (Olson et al., 1980). In contrast, the mRNA for glucose 6-phosphate dehydrogenase begins to increase at 5 hours after feeding fasted rats a high-carbohydrate diet, doubles by 10 h, and reaches its final value after about 24 h (Rudack et al., 1971; Winberry and Holten, 1977). The first detectable increase in protein *synthesis* takes place at about 10 hours, and a new steady-state protein *concentration* is achieved only after 2–3 days. The reason for the pronounced delay prior to induction of the mRNA is not clear, but transcription begins to increase only after 5 h and attains a peak value at 10 h. Induction of malic enzyme is similarly delayed relative to the mRNA after treatment of rats with thyroid hormone or after feeding a high-carbohydrate diet. Due to the wide variation in stability, the predicted delay ranges from a few minutes to more than 60 hours for different pairs of mRNA and protein (Hargrove, 1994).

16.6 Do Exons Coordinate mRNA and Protein Stability?

Note that when the rate of transcription changes, it is possible to have either a rapid increase in concentration of product or a higher concentration of product, but not both (Figure 16.7). Many systems exist in which an mRNA and the encoded protein are both unstable, notably among various protooncogenes and highly regulated enzymes such as ornithine decarboxylase and hepatic enzymes that respond to diet. How could these features have evolved?

Genes are made up of elements that control their expression (by binding *trans*-acting factors), which are mostly located in segments at the 5' end of the gene, and coding segments that are transcribed by RNA polymerase II. The transcribed portions are divided into segments that are retained in the expressed mRNA (*exons*), and those that are removed by RNA processing (intervening sequences or *introns*). The Exon Hypothesis states that functional elements in mRNAs and proteins are often encoded on specific segments of DNA (exons) that can be rearranged so that different genes have common properties. Figure 16.8 compares structural features that are found in mRNAs and other features that are observed in proteins. It is evident that it would be possible for individual exons to encode elements that destabilize (or stabilize) an mRNA and the protein it encodes, albeit by different mechanisms. If time is of the essence in gene expression, this would be one way of ensuring that an mRNA-protein pair is either stable or unstable.

Whether or not coordination of mRNA and protein stability occurred during evolution, there is no question that mRNAs and proteins contain destabilizing elements, and that these elements may be mutated to the advantage of viruses. This is true for the cellular protooncogene, c-*fos*, compared to the viral form, v-*fos*. The last exon in the c-*fos* gene includes coding segments for the carboxylic end of the molecule, as well as the 3', nontranslated portion of the mRNA. The end of the protein contains several structural elements that are commonly found in unstable proteins; these regions contain high amounts of proline, glutamic acid, serine, and threonine, and have been

Exons encode functional units in mRNA and Proteins

Functional Elements in messenger RNA:

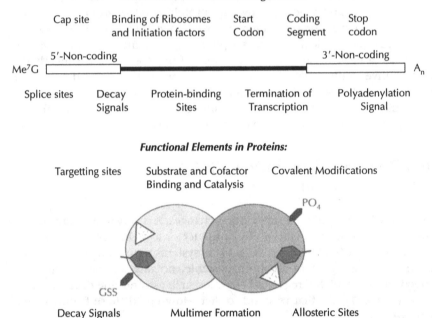

Figure 16.8. Examples of different functions that must be present in mRNAs and the proteins they encode. Many of these functions are thought to be encoded on single exons; it should be evident that exons conferring certain functions on an mRNA must confer different functions on the protein. This mechanism makes it possible to coordinate relative stability of mRNAs and proteins.

termed PEST elements (Rogers et al., 1986). The same exon in the mRNA includes several destabilizing elements that contain UUAUU, represented in Figure 16.8 as U_nA (Van Beveren et al., 1983). These elements are common in cytokine mRNAs, and the mRNA becomes much more stable when units are removed by deletion. Now compare this with the viral oncogene, which has altered the PEST elements by mutation, and has deleted the portion of the exon that contains the U_nA features. The kinetic model predicts that these changes will cause a given rate of transcription to produce a much higher concentration of mRNA and protein, and will produce a switch that cannot be turned off.

16.7 Conclusion

The two-compartment model of gene expression presents an excellent example in which **STELLA®** can serve as a tool for understanding control points in gene expression and possible mechanisms that underlie this control. The theory predicts that it is possible to obtain just as large an increase in concentration of protein by increasing synthesis or by decreasing degradation of either the protein or the mRNA that encodes it. However, the time courses for these processes differ because the degradation rate or stability of the product determines the period needed to achieve a new steady state (by determining the rate of change in concentration). This was evident to David Hoel and colleagues when the dual-compartment model was first developed in 1970, but most investigators at the time were skeptical. In subsequent years, many examples of posttranscriptional control have been reported, and it is clear that nature makes use of every control point possible in situations that require it. This is a case in which theory suggested that such mechanisms should be sought, and **STELLA®** allows any student interested in this topic to perform experiments that demonstrate scenarios now regarded as classic examples of such regulation.

References

Berlin, C.M., and R.T. Schimke. "Influence of turnover rates on responses of enzymes to cortisone." *Mol. Pharmacol.* 1 (1965): 149-156.

DeCrombrugghe, B., S. Busby, and H. Buc. "Cyclic AMP receptor protein: role in transcription activation." *Science* 224 (1984): 831-838.

Goldstein, J.L., and M.S. Brown. "Regulation of the mevalonate pathway." *Nature* 343 (1990): 425-430.

Hargrove, J.L. *Kinetic Modeling of Gene Expression.* Austin: R.G. Landes Co., 1994.

Hargrove, J.L., M.G. Hulsey, and A.O. Summers. "From genotype to phenotype: kinetic modeling of gene expression with **STELLA®**." *BioTechniques* 8 (1993): 654-659.

Hargrove, J.L., and F.H. Schmidt. "The role of mRNA and protein stability in gene expression." *FASEB J.* 3 (1989): 2360-2370.

Lee, K.-L., J.R. Reel, and F.T. Kenney. "Regulation of tyrosine a-ketoglutarate transaminase in rat liver. IX. Studies on the mechanism of hormonal inductions in cultural hepatoma cells." *J. Biol. Chem.* 245 (1970): 5806-5812.

Olson, P.S., E.B. Thompson, and D.K. Granner. "Regulation of hepatoma tissue culture cell tyroxine aminotransferase messenger ribonucleic acid by dexamethasone." *Biochemistry* 19 (1980): 1705-1711.

Parker, R.A., S.J. Miller, and D.M. Gibson. "Phosphorylation of native 97 kDa 3-hydroxy-3-methylglutaryl-coenzyme A reductase from rat liver. Impact on activity and degradation of the enzyme." *J. Biol. Chem.* 254 (1989): 4877-4887.

Pilkis, S.J., M.R. El-Maghrabi, and T.H. Claus. "Hormonal regulation of hepatic gluconeogenesis and glycolysis." *Annu. Rev. Biochem.* 57 (1988): 755-784.

Rogers, S., R. Wells, and M. Rechsteiner. "Amino acid sequences common to rapidly degraded proteins: the PEST hypothesis." *Science* 234 (1986): 364-368.

Rudack, D., E.M. Chisholm, and D. Holten. "Rat liver glucose 6-phosphate dehydrogenase." *J. Biol. Chem.* 246 (1971): 1249-1254.

Schimke, R.T. "Control of enzyme levels in mammalian tissues." *Adv. Enzymol.* 37 (1973): 135-187.

Van Beveren, C., F. van Straaten, T. Curran, R. Muller, and I.M. Verma. "Analysis of FBJ-MuSV provirus and c-fos (mouse) gene reveals that viral and cellular fos gene products have different carboxy termini." *Cell* 32 (1983): 1241-1255.

Winberry, L., and D. Holten, "Rat liver glucose-6-P dehydrogenase." *J. Biol. Chem.* 252 (1977): 7796-7801.

Yagil, G. "Quantitative aspects of protein induction." *Curr. Top. Cell. Reg.* 9 (1975): 183-286.

17

From Genotype to Phenotype

Between the characters that are used by the geneticist and the genes that
his theory postulates lies the whole field of embryonic development,
where the properties implicit in the genes become explicit in the proto-
plasm of the cells. Here we appear to approach a physiological problem,
but one that is new and strange to the classical physiology of the schools.

Thomas Hunt Morgan,
Nobel Lectures in Molecular Biology, 1933–1975,
New York: Elsevier North-Holland, 1965

17.1 Toward a Comprehensive Model of Gene Expression

The idea that rates of degradation influence the time course of approach to
new, steady-state concentrations applies to mRNAs and enzymes, but the
two-compartment model is incomplete. It does not account for the processes
by which transcription is activated, which frequently involve the regulation
of DNA-binding proteins by ligands or by covalent modification. Also it does
not include the processes involved in mRNA processing and transport, or the
impact of the completed protein on cellular metabolism. To be useful to the
molecular biologist, a simulation tool should be able to describe any intermediate that could potentially affect the outcome of an experiment, and this
requires information about each rate of formation, transfer, or elimination
(Hargrove, 1993). In the sequential reactions of gene expression, the processing or transfer of each intermediate not only affects the yield, but also introduces a delay that is proportional to the half-time of each step. Many molecular biologists need to make predictions concerning the effects of altered
transcription, nuclear mRNA metabolism, and translational or posttransla-
tional controls.

Although each reaction in the process of gene expression is characterized by an individual rate or half-time for completion, the system cannot
attain equilibrium until each process has attained a constant rate. The interval required for the system to equilibrate may be defined as the *relaxation
time* (Fersht, 1985; Almagor and Paigen, 1988), and experimental data that
deal with induced mRNA or protein synthesis reflect this aggregate property.
This chapter addresses this concept, shows how to go about assembling a
basic model, and demonstrates how kinetic modeling may be used to make
predictions and integrate primary data concerning more inclusive models. In
kinetic terms, a strictly compartmental model of gene expression would
be called linear. However, the process is controlled by saturable binding of

factors that are involved in each control point, so the outcome may be distinctly nonlinear.

17.2 Creating a Computer Program to Simulate Gene Expression

A general 'model of gene expression should include at least three components: (1) the phase of transcriptional activation, which is a nonlinear process; (2) the central steps of nuclear mRNA processing, cytoplasmic mRNA translation, and protein accumulation; and (3) functional effects of the protein, which are nonlinear for enzymes. The first step in compartmental analysis is to create a diagram of the system of interest, and Figure 17.1 illustrates a schema that begins with transcription and includes the primary RNA transcript, processed nuclear mRNA, cytoplasmic mRNA, and protein. As represented in the **STELLA®** modeling system, each constituent that accumulates is represented as a rectangle (reservoir or stock), and each process that generates the constituent is represented by a flow into the reservoir, with each fractional transfer rate designated λ_i. Precedents for modeling the activation of transcription, translational controls, and the impact of proteins on metabolic control will be discussed at the end of the chapter. Let us start with a model that includes all nuclear mRNA (designated N), with cytoplasmic mRNA shown as R, and the protein called P (Figure 17.2).

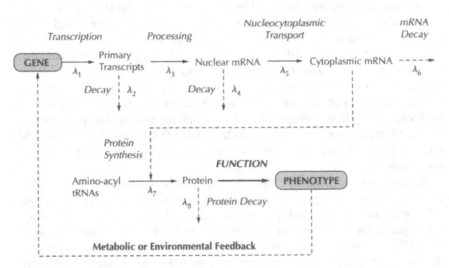

FIGURE 17.1. A kinetic model of eukaryotic gene expression including nuclear decay and nucleocytoplasmic transport is shown. A compartmental model may be created with each intermediate represented as a pool or stock in a **STELLA®** diagram. Each inflow or outflow is designated by a characteristic rate parameter, λ_i. Provision for metabolic control is indicated by the feedback loop by which an enzyme or protein alters a specific phenotypic trait, which then acts via a signal to modify gene expression.

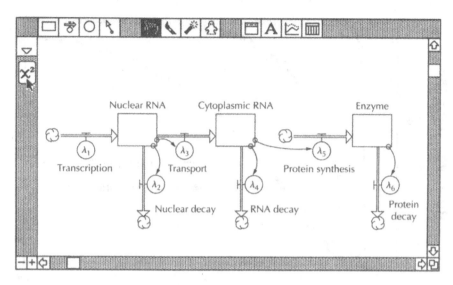

FIGURE 17.2. A simplified three-compartment model of mRNA and protein metabolism was produced using **STELLA®**. Rates of synthesis and decay or interconversion were entered by opening dialog boxes associated with each open symbol in the diagram. Simulations performed with this model predict a rapid accumulation of nuclear mRNA, followed by sigmoidal increases in cytpoplasmic mRNA and protein.

When flow out of one compartment into another is proportional to concentration, the rate of formation of each product is equal to the concentration of the previous intermediate multiplied by the fractional amount that enters the next compartment (C_{i-1} times λ_{i-1}). Similarly, each outflow equals the concentration in the pool of interest times the appropriate transfer rate (C_i times λ_i). Material in each of these compartments is predicted to accumulate and decay according to the elimination rates that apply, with the net effect that each compartment introduces a delay in the flow of genetic information. Let us designate the zero-order rate of formation of nuclear mRNA as λ_1, the rate of transcription; the rate of formation of cytoplasmic mRNA then equals $\lambda_3 N$, and the rate of formation of protein equals $\lambda_5 R$. Based on Figure 17.2, the following equations can be written to express the concentration of each product at steady-state:

$$N = \lambda_1/(\lambda_2 + \lambda_3) \tag{17.1}$$

$$R = \lambda_3 N/\lambda_4 = [\lambda_1\lambda_3]/[(\lambda_2 + \lambda_3)(\lambda_4)] \tag{17.2}$$

$$P = \lambda_5 R/\lambda_6 = [\lambda_1\lambda_3\lambda_5]/[(\lambda_2 + \lambda_3)(\lambda_4\,\lambda_6)] \tag{17.3}$$

Frankly, these details make the problem seem harder to understand than it is. Please refer to the **STELLA®** diagram to help simplify the discussion. All we are saying is that each compartment has at least one inflow and outflow rate, and that it is possible to calculate amounts of material in each compartment if we know these values.

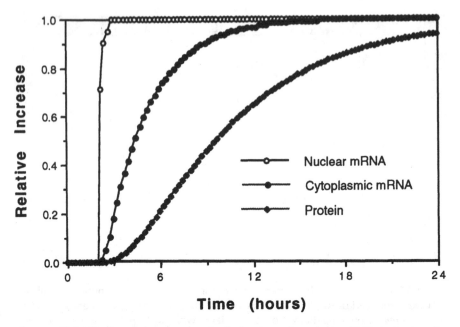

FIGURE 17.3. Predicted time course for induction of nuclear mRNA, cytoplasmic mRNA, and protein after the rate of transcription increases. The kinetic model shown in Figure 17.2 was used with rates of decay corresponding to the following half-lives: nuclear mRNA, 0.33h (20 min); cytoplasmic mRNA, 2 h; protein, 4 h.

According to the diagram, the newly synthesized mRNA enters a nuclear pool at a rate called λ_1 molecules per cell per hour, and is either degraded (rate = λ_2 h^{-1}), or is processed and transported to the cytoplasm (rate = λ_3 h^{-1}). For example, if the rate of production equals 1 molecule per cell per hour, and the half-time for each transfer is 42 min (0.693 h), then the fractional transfer rates equal 0.693/0.693, or 1 h^{-1}, and the concentration of nuclear mRNA equals 1/(1 + 1), or 0.5 molecules cell^{-1} h^{-1}. The rate at which mRNA will appear in the cytoplasm equals the concentration times transfer rate, or 0.5 times 1.0, so that the rate of appearance in the cytoplasm equals 0.5 molecules h^{-1}. To calculate mRNA concentration at steady state, the half-life must be known. If the half-life equals 2 h, then the fractional rate of disappearance is 0.693/2, or 0.3465 h^{-1}. This means that the concentration of mRNA at steady state will equal 0.5 molecules per cell per hour divided by 0.3465 h^{-1}, or 1.44 molecules per cell. Similarly, if 100 molecules of protein are made per molecule of mRNA per hour, and the protein has a half-life of 4 h, then the fractional turnover rate of the protein is 0.693/4 h, or 0.1733 h^{-1}, and the protein concentration equals 1.44(100/0.1733), or 831 molecules per cell.

It is better just to enter the numbers into the **STELLA**® model and see what the outcome is. Figure 17.3 shows such a result. The nuclear mRNA has a short half-life and reaches a new level quickly after the rate of synthesis increases. This intermediate introduces a delay in the increase in the cytoplasmic mRNA, and protein synthesis begins to increase as the cytoplasmic message increases. Based on the half-lives assumed in the simulation, this protein will require 24 h to come to a new equilibrium.

In terms of control of gene expression, an important conclusion may be derived from the steady-state equations. The equations predict that only changes in the rate of transcription (λ_1) and the rate of decay of cytoplasmic mRNA (λ_4) would produce equivalent effects on mRNA concentration at the steady state. Changes in the terms for nuclear decay or nucleocytoplasmic transport would produce smaller effects because these two processes compete for the same nuclear precursor; if the model included an extra term that represented processing of the primary transcript, the same conclusion would apply. The competition is expressed quantitatively by representing these rates as additive terms in the denominator of Eqs. 17.2 and 17.3. An increase in nucleocytoplasmic transport without an increase in transcription would deplete the pool of nuclear mRNA, and therefore would not proportionally increase the amount of cytoplasmic mRNA. For instance, if each parameter had a value of 1, then cytoplasmic mRNA (R) would equal 0.5, and a tenfold increase in rate of transcription or decrease in decay of cytoplasmic mRNA would produce a tenfold increase in RNA. A hand-held calculator can be used to show that a tenfold increase in the rate of processing or export, or increased stabilization of nuclear precursor, would increase the concentration by less than twofold (from 0.5 to 0.9). A larger effect would be observed, however, in the unusual case in which the majority of the transcript was normally degraded in the nucleus. Then nuclear stabilization would increase efflux to a greater extent than in the example just given; only if the cell could choose between complete degradation and no degradation would the effect be equivalent to a corresponding change in transcription. This suggests that control mechanisms that enhance transcription, stabilize cytoplasmic mRNA, or control translation are all more efficient than mechanisms that act on processing or nucleocytoplasmic transport of mature mRNA.

A change in the half-life of the cytoplasmic mRNA would be expected to produce a large effect on the time course of induction because λ_4 represents the time constant for change in concentration of translated mRNA. However, the time scale of change for mRNA precursor is expected to be short, as dictated by $\lambda_2 + \lambda_3$. For most eukaryotic mRNAs, the rate of cytoplasmic decay is much slower than the time constants for the nuclear processes, and the change in concentration of mRNA in the cytoplasm would be expected to be influenced relatively little by the nuclear events.

In some cases, a molecule may be depleted from a pool by more than one route. This is true for nuclear mRNA, which may be degraded or exported to the cytoplasm, and for secretory proteins, which may be degraded or secreted. If the outflows are proportional to concentration, then flow through each efflux pathway may be calculated by multiplying the concentration of product by the fractional rate corresponding to each pathway. The effective rate at which the pool is depleted is then equal to the sum of the rate parameters. For example, in Figure 17.1, the effective rate of depletion of the nuclear mRNA equals $\lambda_2 + \lambda_3$.

17.3 Translational Control

Numerous examples of translational control mechanisms are now known, for example, the regulation of ferritin synthesis by the iron regulatory element (Hentze et al., 1989; Munro, 1990), and the control of ornithine decarboxyl-ase by the presence of polyamines (Persson et al., 1988; Manzella and Black-shear, 1990). How can a simulation program be modified to include this pro-cess? STELLA® software programs make this very easy to do. For example, the basic diagram may be modified as shown in Figure 17.4, which shows that cytoplasmic mRNA exists in translationally active and inactive forms, designated R and U, respectively. Steady-state equations for any diagram in which a component flows reversibly between two compartments can be ob-tained by setting up simultaneous equations, solving for one component in terms of the others, and substituting to obtain the correct values for use in the computer program.

Recall that net flux equals zero in the steady state, and that the quantity in each compartment is a function of the rates of influx and efflux. The rate of protein synthesis will still be a function of the abundance of translated mRNA, as discussed in the previous section, and may be ignored. Note that the amount of mRNA that initiates protein synthesis, R, is formed not only from newly synthesized RNA, but also from the untranslated pool, as repre-

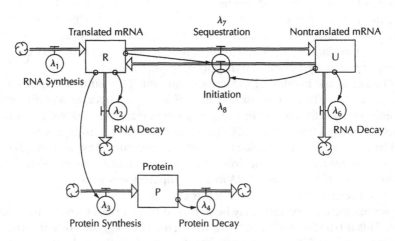

FIGURE 17.4 A model including posttranscriptional control is easily constructed with STELLA® by adding an extra compartment for untranslated mRNA and a means of controlling the rate of interconversion. The consequence of sequestering the mRNA is that it need not be synthesized and can begin protein synthesis very rapidly, thus bypassing the delay imposed by accumulation controlled by high RNA stability.

sented by the two parameters, λ_1 and λ_8, and is depleted not only by mRNA degradation (λ_2), but also by sequestration into the nontranslated pool (λ_7). The mass balance equation for synthesis and degradation of the translated mRNA is: $\lambda_1 + \lambda_8 U = R \lambda_2 + R \lambda_7$. Therefore, the steady-state equations can be written as follows:

$$R = \frac{\lambda_1 + \lambda_8 U}{(\lambda_2 + \lambda_7)} \tag{17.4}$$

$$U = \frac{\lambda_7 R}{\lambda_6 + \lambda_8} \tag{17.5}$$

The initial values for R and U can be solved by substituting the value of U from Eq. 17.5 into Eq. 17.4. To make this easier, define constants $A = \lambda_1/(\lambda_2 + \lambda_7)$, $B = \lambda_8/(\lambda_2 + \lambda_7)$, and $C = \lambda_7/(\lambda_6 + \lambda_8)$. Then

$$U = CR \tag{17.6}$$

$$R = A + BU = A + BCR \tag{17.7}$$

Therefore,

$$R = A/(1 - BC) \tag{17.8}$$

The steady-state values for the translated pool can be obtained from Eq. 17.8, and the value for the nontranslated pool of mRNA can be calculated from Eq. 17.6. Notice that for any set of kinetic parameters, at steady state the relative amount of RNA in the two pools is $U/R = C$. If the protein were interconverted between active and inactive forms, the same method would be used to solve for the concentrations of the two forms. After the steady-state values have been calculated, the computer program may be used to simulate effects of changing various controls.

In practice, the relative quantities are regulated in a manner that depends on the binding of *cis*-acting factors or other control mechanisms. The software permits this to be modeled as a switch that is either on or off (presence or absence of hemin or free iron in the example of ferritin), or by adding a separate module that represents the concentration of iron and of the iron-dependent regulatory protein. This is easy to do with the simulation package, but the user must be cautious of one foible that is intrinsic to programs based on finite difference equations: If kinetic parameters are expressed as fractional exchange rates per hour, and a control can occur within minutes, then the default calculation interval must be set to a shorter value. As a rule of thumb, the calculation interval should never exceed half of the shortest time constant in the model, or else the system will oscillate erratically. For example, the default calculation interval in the **STELLA®** program is 0.25 time units, or hours in the present case. If the translational mechanism has a half-time of 5 minutes (0.083 h), then the calculation interval should be changed to a value less than or equal to 0.04 h.

The importance of translational controls lies in their ability to permit rapid synthesis of new protein without the delays imposed by the other rate parameters. If the half-life of mRNA in the cytoplasm was 2 h, then first-order kinetics suggests that a transcriptionally active stimulus would require at least 2 h to achieve half-maximal rates of protein synthesis, and about 10 h to attain a new rate of synthesis. In contrast, preformed mRNA could become translationally active in a matter of minutes, which is very important in preventing iron-dependent oxidative damage to the cell. The only other alternatives for rapid acquisition of extra capacity would include the use of preformed protein, which is not always feasible due to the many competing needs of the cell, or an increase in the amount of protein synthesized per mRNA molecule.

17.4 The Activation of Transcription

The original kinetic models of enzyme induction considered only three elements: the concentration of an enzyme, its rate of synthesis, and its rate of elimination. Now that the processes by which transcription is controlled are much better understood, it is legitimate to ask how to generate a realistic model that is accessible to scientists who have a basic grasp of kinetic principles. The aim of this section is to discuss the simplest possible way of accomplishing this task, recognizing that although every gene is controlled by numerous factors, most experimentalists only study one or two at a time.

The idea that enzyme induction could be modeled as "zero-order synthesis, first-order decay" was known to be an oversimplification when first proposed because it omitted details of the processes involved in transcription and translation. Technically, that model is only correct when the rate of synthesis changes instantaneously from one level to another. The model worked relatively well because mRNAs and proteins in mammals are relatively stable once synthesized, with half-lives that usually exceed 1–2 h and frequently are much longer than this. For instance, the average half-life of mRNA in liver has been reported to range from about 5–20 h (Moore et al., 1980), and the average half-life of hepatic protein is in the range of 3–5 days (Schimke, 1973; Hargrove et al., 1991). Because the time course for approach to steady state is limited by the slowest rate parameters in effect, the short periods needed to activate transcription can frequently be ignored. For many mammalian genes, transcription can be activated to its maximum extent within 15–30 min. This means that the half-time for transcriptional activation typically represents about 5% of the time needed to achieve a half-maximal increase in mRNA, and less than 1% of the time needed for half-maximal increase in protein. Therefore, the idea that induction can be modeled with the rate of synthesis represented as zero order frequently is a very reasonable assumption. Still, there are numerous examples in the literature in which this model is inadequate because the accumulation curve for mRNA is sigmoidal, and often is associated with a delay that is not predicted from the basic model. Because the events that produce the delay and the sigmoidal shape are often central to understanding the mechanism of control, it is important to discuss how these components can be incorporated into a kinetic model.

17.5 Kinetics of Transcriptional Activation

If the binding of proteins to DNA only involved one target DNA sequence and one site on the DNA-binding protein, then the reaction would be expected to follow a binding curve predicted by Michaelis-Menten kinetics, and a plot of the fraction of available binding sites versus concentration of the protein would be expected to be hyperbolic. In practice, it seems unlikely that such simple kinetics will be common for several reasons:

1. Many DNA-binding proteins are dimers or multimers, and the dimerization reaction causes the binding curve to have a sigmoidal shape, whether the curve is expressed in terms of protein concentration or in terms of the time course of the reaction.
2. Many DNA-binding proteins are regulated by ligands or by covalent modification, and the sequence of ligand binding followed by binding of the complex to DNA produces a sigmoidal curve.
3. Multiple binding sites for specific regulatory proteins exist in promoter elements, and this creates the possibility for allosteric effects, cooperativity, and threshold behavior.
4. Cooperative interactions occur not only between DNA-binding proteins and RNA polymerase, but also with other regulatory factors.

Many DNA-binding proteins are activated or inactivated either as a result of ligand binding, or because they are controlled directly or indirectly by regulatory enzymes. This means that transcriptional controls may be simulated by using well-established principles for analyzing the interactions of ligands with their receptors, or of substrates with enzymes (Taylor and Insel, 1990; Levitski, 1984; Koshland, 1987). After a DNA-binding protein has been activated, its binding to regulatory elements in DNA also has a kinetic aspect that must be considered.

17.6 Rates of Association Between Ligands and Their Receptors

The minimum number of processes that are necessary to simulate effects of DNA-binding proteins on transcription are two: initial binding of a ligand to its specific receptor, and subsequent binding of the ligand-receptor complex to DNA (or dissociation of the complex from DNA). Examples of this kind of process include the binding of steroidal hormones, second messengers, metabolites, and metal ions to their receptors.

In the case in which a single ligand binds to one site on a receptor molecule, the rate of association and dissociation can be defined from the following equation (Taylor and Insel, 1990):

$$L + R \underset{k_2}{\overset{k_1}{\rightleftharpoons}} LR \xrightarrow{k_3} LR^*$$

The rate of association equals $k_1[L][R]$, in which k_1 has units, $M^{-1}time^{-1}$, or liters*mole^{-1} time^{-1}. *Note that if k_1 is multiplied by the ligand concentration, the rate of association ($k_1[L]$) now has units of time^{-1}.* In other words, the time dependency of the response is a function of k_1 and ligand concentration, as has been shown clearly for binding of dexamethasome to the glucocorticoid receptor (Baxter and Tomkins, 1971). The rate of dissociation is first order; it equals $k_2[LR]$, and k_2 has units of time^{-1}. Although this is the most basic formulation, in some cases the LR complex changes conformation and becomes active for binding, in which case the active complex, LR*, is regarded as a distinct entity with a unique rate constant for formation. For the time being, we will ignore the transition to the active state, and ask only how the concentration of the complex varies with time.

If only the association reaction is considered, then the rates of association and dissociation are equal at equilibrium,

$$k_1[L][R] = k_2[LR] \tag{17.9}$$

This is related to the apparent equilibrium association and dissociation constants as follows:

$$K_D = \frac{[L][R]}{[LR]} = \frac{k_2}{k_1} \tag{17.10}$$

Therefore, the dissociation constant at equilibrium can be defined either as a ratio of concentrations or as a ratio of dissociation and association rate constants. In the kinetic analysis, the rate constants are more important because they include the dimension of time. Notice that the K_D has units of concentration, and represents the concentration of ligand at which a half-maximal binding response is observed.

How does the LR complex concentration change with time?

$$\frac{d[LR]}{dt} = k_1[L][R] - k_2[LR] \tag{17.11}$$

When the concentration of ligand is much greater than R, which is often the case, the rate of approach to equilibrium equals

$$\ln \frac{[LR]_{eq}}{([LR]_{eq} - [LR]_t)} = (k_1[L] + k_2)t \tag{17.12}$$

In other words, the rate at which the LR complex forms depends on both the forward and reverse kinetic constants and on the concentration of ligand. A plot of the function on the left side of this equation over time gives a hyperbolic curve when the ligand is present at high concentration, and the logarithmic function is a straight line.

An alternative way of expressing this information in terms of ligand concentration is to consider the binding at any concentration of ligand as a fraction of the total possible binding in the tissue. This can be expressed similarly to the Michaelis-Menten expression as follows:

$$[LR] = \frac{[L][R]_{total}}{K_D + [L]} = \frac{[L][LR]_{max}}{K_D + [L]} \qquad (17.13)$$

because the maximum amount of complex is equal to the total receptor present in the system. As with the Michaelis-Menten equation, this represents a rectangular hyperbola that is half maximal when the ligand concentration equals the value of the K_D. This is actually a single solution to a more general equation in which there is cooperativity among binding sites for multimeric units. In this case, the Hill equation represents a more general solution (Koshland, 1987). Basically, it represents the fraction of ligand that is bound in an active complex at any concentration, multiplied by the maximal possible complex:

$$[LR] = \frac{[L]^n[LR]_{max}}{[L_{0.5}]^n + [L]^n} \qquad (17.15)$$

where the concentration of ligand giving half-maximal response is defined as $L_{0.5}$, and equals the K_D. The exponent n can have any positive value that does not exceed the number of binding sites; it need not be an integer. The fractional amount of LR present at any time is equal to:

$$\frac{[LR]}{([LR]_{max} - [LR])} = \frac{[L]^n}{L_{0.5}^n} \qquad (17.16)$$

Strickland and Loeb (1981) employed a similar kind of analysis to link the generation of second messengers to receptor saturation and subsequent biological response.

17.7 Delays and the Idea of Relaxation Time

Just as musical compositions have time signatures, every complex biological process has duration and requires predictable periods to approach a new equilibrium. The aggregate rate for the process depends on the sum of the rates of the constituent processes, and some contribute more to the overall rate than do others. Experimentally, the process of identifying the contributions of each subprocess yields important information about mechanisms. This idea has received theoretical treatment in the concept of *relaxation times*, by which the biochemist is able to identify all those processes that contribute to the time needed for a reaction to approach its equilibrium (Fersht, 1985). In this technique, a change is introduced and the time required for the reaction to seek a new equilibrium is determined. The system proceeds to this equilibrium via a series of relaxation times, t (= the reciprocal of the rate constant[s]). Effectively, all the identifiable rates are added together to achieve a global value that depends on the sum of its parts. And because each step in gene expression has a magnitude, the initial value may be amplified or attenuated, thus introducing the idea of scaling into the picture. The hierarchical organization of the human and animal body makes this last idea critical, for the processes initiated by one or two genes per cell must be magnified to change a property, or phenotype, of the whole organism.

Often, gene expression is induced by addition of a high concentration of ligand such as a steroidal hormone of a cyclic AMP analog, and the contribution of the ligand concentration to time course is ignored. Even under these circumstances, it is common to find that the rate of transcription from different genes, as well as rates of mRNA and protein accumulation, differ. What mechanisms can account for these differences? Almagor and Paigen (1988) have modeled induced gene expression as a function of binding of transacting factors to promoter elements, and concluded that the minimum expression needed to simulate the time course includes three separate functions, namely, the rate of association and dissociation of the binding protein from the DNA, the half-life of the mRNA, and the half-life of the protein. In addition, their model included the supposition that several binding sites for one factor may exist on a single gene. The model assumed that the activated receptor complex bound to a DNA element designated s, and converted it to a DNA-receptor complex designated s*.

$$s \underset{k_2}{\overset{k_1}{\rightleftharpoons}} s^*$$

where both rate constants have units of time^{-1}. When there are n different binding sites for the receptor, the fraction of the genes with a given number of bound receptors after addition of the inducing agent is expressed as follows:

$$X_t = \frac{k_1}{(k_1 + k_2)}(1 - e^{-(k_1+k_2)t}) \qquad (17.17)$$

The sum of the two rate constants, $k_1 + k_2$, was defined as the relaxation time for binding, and was designated k^0. In other words, the rate at which a specific fraction of DNA-binding elements was occupied was determined by the sum of the two rate constants. The average occupancy of the sites in a population of cells is then equal to the number of sites available on each promoter (N) times X_t, and the rate at which this occupancy is attained is a simple exponential function.

To duplicate the effects of this mechanism on gene expression, it is necessary to add to this the equations for accumulation of mRNA and protein, so that the result has the following form:

$$P_t = P_{ss} - Ae^{-k_ot} - Be^{-k_rt} - Ce^{-kpt} \qquad (17.18)$$

where A, B, and C are constants related to the kinetic constants, and P_{ss} equals the mRNA concentration at steady state multiplied by the rate of pro-

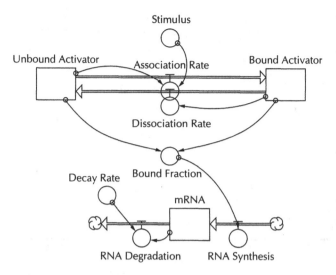

FIGURE 17.5. The activation of DNA-binding proteins also adds a delay to initiation of transcription. The rate of activation depends not only on the rate at which the active complex binds DNA, but also on the rate at which the DNA-binding factor becomes competent to bind.

tein synthesis per unit of mRNA, and divided by the rate of protein degradation. Effectively, this is the sum of three exponential curves, and has a sigmoidal shape. This seemingly difficult equation represents nothing more than a three-compartment **STELLA**® model!

The main conclusion based on the consideration of ligand-dependent activation of a receptor that binds to DNA and the subsequent binding of the complex to DNA, is that effective simulation of gene activation must include at least one saturation curve, and this introduces a time dimension into the activation process that in some cases amounts to several hours. The temporal response of the entire process includes the activation phase and period needed for mRNA and protein to accumulate. Other mechanisms that can also produce delays include cooperative binding at multiple sites, thresholds for activation of transcription, and physical changes in the initiation complex (Yamamoto and Alberts, 1975; Mulvihill and Palmiter, 1977; Palmiter et al., 1976).

The description of how ligands and receptors interact provides a means of using a computer program such as **STELLA**® to link the binding of a transacting factor to a gene, altered transcription, and accumulation of mRNA or protein products as a function of time and ligand concentration. In the example diagrammed in Figure 17.5, the rate of mRNA synthesis was assumed to be a function of the fraction of a regulatory molecule that was bound to the promoter. As shown in Figure 17.6, this introduces a delay compared to the single-compartment model, so the predicted mRNA response curve is

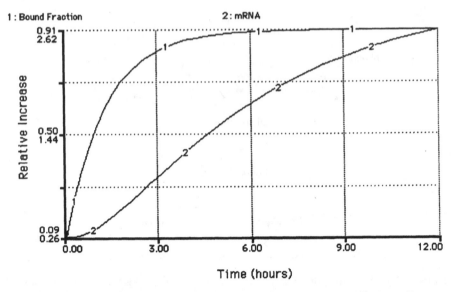

FIGURE 17.6. Simulation of transcriptional activation by a *trans*-acting factor indicates the delay imposed by the time required to achieve a maximal bound fraction. The consequence is to cause a delay in mRNA induction, which produces a sigmoidal accumulation curve for the mRNA. Plainly, the consequence is that the rate of appearance of mRNA in the cytoplasm of a eukaryotic cell depends on the rate at which transcription is activated, as well as on the rates of nuclear mRNA processing and transport.

sigmoidal even if nuclear processing is ignored, as suggested by Almagor and Paigen (1988). The length of the delay depends on the rate at which the trans-acting factor was activated, and would be negligible if the activation was very rapid compared to the kinetics of mRNA accumulation. Any of these equations may used in modeling programs to achieve outcomes that simulate experimental data.

17.8 The Law of Diminishing Returns

Finally, if one is considering a protein that is an enzyme, it is necessary to ask what effect a change in enzyme concentration has on flux of a substrate through a pathway. The change in flux as a function of enzyme activity cannot be predicted in advance. It depends on characteristics of the enzymes in question, the quantities of each in the cell under initial conditions, and their kinetic parameters such as affinities, half-maximal velocities, and influence of products and allosteric modulators (Westerhoff et al., 1984; Liao and Delgado, 1993). In well-studied systems, the measured responses can be plotted as hyperbolic curves, with small changes from the initial values producing relatively large effects on flux. However, as enzyme concentrations increase, the effect on flux becomes smaller and smaller, approaching an asymptote of zero.

Hepatic metabolism of amino acids is affected by the activity of enzymes that are induced by dietary protein and by hormones such as cortisol and glucagon. In situations in which diets contain moderate levels of protein (10–20% by weight), the enzyme activities are low, but the activities increase by five- to tenfold as dietary protein is increased, and by similar values when animals are treated with glucocorticoid hormones that mimic the action of cortisol or corticosterone. In a study of aromatic amino acid metabolism in brain and liver, Salter and colleagues (1986) found that metabolism of tryptophan and tyrosine was affected both by the rates of transport into cells and by the concentrations of specific enzymes in the metabolic pathways by which these amino acids were metabolized. They measured the flux as a function of the activities of tryptophan oxygenase and tyrosine aminotransferase. The data for the change in flux as a function of enzyme quantity fit hyperbolic curves of the following form:

$$J = \frac{J_{max} V}{(V + k)} \qquad (17.19)$$

where J is observed flux under experimental conditions, J_{max} is predicted flux at infinite enzyme velocity, V is assayable maximum activity for the enzyme in question when assayed *in vitro*, and k is a constant defined as the value of V when $J = J_{max}/2$.

If one intends to model the effects of altered transcription on the ability to metabolize an amino acid, the consequences of this theory are important. First, it is clear that a specific increase in transcription may eventually produce an equivalent increase in enzyme concentration, and therefore an increase in the assayable enzyme activity. This does not mean that flux will change in proportion. Instead, the Law of Diminishing Returns holds, because the more the enzyme activity increases, the less it controls the flux through a particular metabolic pathway. Eventually, the rate of metabolism will be limited by substrate availability, by the rate of another enzyme in the pathway, or by accumulation of product. In fact, this has a very practical implication for the mammalian cell. For enzymes that control the rate of reactions but that are present at significant amounts in the "basal" condition, small increases in concentration will effectively increase the rate of metabolism. However, an increase in concentration beyond a certain level is ineffective. In the data obtained by Salter and colleagues (1986), the limiting enzymes controlled about 70% of flux through the pathway in the basal state, but only about 10–20% of flux in the induced state. If the point of regulation is to achieve altered rates of metabolism, plainly it is not effective in cases such as this to increase transcription more than tenfold. This is important for cellular economy, because cell size and resources for protein synthesis are limited, and there are competing demands for enzymes in other pathways.

17.9 Conclusions

Most messenger RNAs and proteins in mammalian cells have half-lives that range from several hours to several days. For this reason, changes in many proteins occur only slowly. It is not uncommon to find that changes in mRNA production are observed during the course of a day, but that the protein concentration never varies significantly. The kineticist would say that the oscillation is damped; the flow of information from gene to protein is not immediate. Many regulatory mRNAs and proteins, however, necessarily are unstable and therefore are able to respond quickly to any stimulus that requires their concentrations to change.

In contrast to the situation in vertebrate animals, messenger RNAs in microbes are very unstable, typically having half-lives of 1–3 *minutes*. This confers two kinds of advantage: the microbe is able to alter its enzyme repertoire quickly to adjust to new metabolic demands, and the microbial molecular biologist is able to finish his experiments in less than a day! Pity the graduate student who has been assigned to study stable mRNAs and enzymes, for his or her experimental system may require a week to reach a steady state! More importantly, however, consider the fact that each cell has limited volume, with constant demands for metabolic adaptation. In the words of Helen Tepperman and colleagues (1963), "the business of the liver is eternally unfinished."

References

Almagor, H., and K. Paigen. "The chemical kinetics of induced gene expression. Activation of transcription by noncooperative binding of multiple regulatory molecules." *Biochem.* 27 (1988): 2094–2102.

Baxter, J.D., and G.M. Tomkins. "Specific cytoplasmic glucocorticoid hormone receptors in hepatoma tissue culture cells." *Proc. Natl. Acad. Sci. USA* 68 (1971): 932–937.

Fersht, A. *Enzyme structure and mechanism*. New York: W.H. Freeman and Co., 1985.

Hargrove, J.L. "Microcomputer assisted kinetic modeling of mammalian gene expression." *FASEB J.* 7 (1993): 1163–1170.

Hargrove, J.L., M. Hulsey, and E.G. Beale. "The kinetics of mammalian gene expression." *BioEssays* 13 (1991): 667–674.

Hentze, M.W., T.R. Rouault, J.B. Harford, and R.D. Klausner, "Oxidation-reduction and the molecular mechanism of a regulatory RNA-protein interaction." *Science* 244 (1989): 357–359.

Koshland, D.E., Jr. "Switches, thresholds, and ultrasensitivity." *TIBS* 12 (1987): 225–229.

Levitzki, Alexander. In: *Receptors: A Quantitative Approach*. Menlo Park, Cal.: Benjamin/Cummings Publishing Co., 1984.

Liao, J.C., and J. Delgado. "Advances in metabolic control analysis." *Biotechnol. Prog.* 9 (1993): 221–233.

Manzella, J.M., and P.J. Blackshear. "Regulation of rat ornithine decarboxylase mRNA translation by its 5'-untranslated region." *J. Biol. Chem.* 265 (1990): 11817–11822.

Moore, R.E., T.L. Goldsworthy, and H.C. Pitot. "Turnover of 3'-polyadenylate-containing RNA in livers from aged, partially hepatectomized, neonatal, and Morris 5123C hepatoma-bearing rats." *Cancer Res.* 40 (1980): 1449–1457

Mulvihill, E.R., and R.D. Palmiter. "Relationship of nuclear estrogen receptor levels to induction of ovalbumin and conalbumin mRNA in chick oviduct." *J. Biol. Chem.* 252 (1977): 2060–2068.

Munro, H.N. "Iron regulation of ferritin gene expression." *J. Cell. Biochem.* 44 (1990): 107–115.

Palmiter, R.D., P.B. Moore, E.R. Mulvihill, and S. Emtage, "A significant lag in the induction of ovalbumin messenger RNA by steroid hormones: a receptor translocation hypothesis." *Cell* 8 (1976): 557–572.

Persson, L., I. Hom, and O. Heby. "Regulation of ornithine decarboxylase mRNA translation by polyamines." *J. Biol. Chem.* 263 (1988): 3528–3533.

Salter, M., R.G. Knowles, and C.I. Pogson. "Quantification of the importance of individual steps in the control of aromatic amino acid metabolism." *Biochem. J.* 234 (1986): 635–647.

Schimke, R.T. "Control of enzyme levels in mammalian tissues." *Adv. Enzymol.* 37 (1973): 135–187.

Strickland, S., and J.N. Loeb. "Obligatory separation of hormone binding and hormone action." *Proc. Nat. Acad. Sci. USA* 78 (1981): 1366–1370.

Taylor, P., and P.A. Insel. "Molecular basis of pharmacologic selectivity." In: *Principles of Drug Action, The Basis of Pharmacology*, Pratt, W.B., and P. Taylor, eds. New York: Churchill Livingstone, 1990.

Tepperman, H., J. Tepperman, J. Pownall, and A. Branch. "On the response of hepatic glucose 6-phosphate dehydrogenase activity to changes in diet composition and food intake pattern." *Adv. Enzyme Reg.* 1 (1963): 121–136.

Westerhoff, H.V., A.K. Groen, and R.J.A. Wanders. "Modern theories of metabolic control and their application." *Bioscience Reports* 4 (1984): 1–22.

Yamamoto, K.R., and B. Alberts. "The interaction of estradiol receptor protein with the genome: an argument for the existence of undetected specific sites." *Cell*, 4 (1975): 301–310.

18

The Plateau Principle: A Key to Biological System Dynamics

I worked out the equations and came up with the surprising and satisfyingly simple result that the halftime for a shift in steady state protein concentration depended only on the rate of protein degradation, regardless of the rate of synthesis. It was Avram (Goldstein) who saw the general applicability of this approach. . . .

Dr. Dora B. Goldstein

There may have been a time when only the clinical pharmacologist and the experimental nutrition scientist truly needed to understand the dynamics of materials introduced into the human body. All that changed with the passage of the Dietary Supplement Health and Education Act of 1994, about which information is available from the U.S. Food and Drug Administration's Center for Food Safety and Applied Nutrition (http://www.cfsan.fda.gov). It is now possible to purchase powerful hormones and plant-derived substances over the counter and without prescription from any drugstore in the United States. The freedom is refreshing, but let the buyer beware; our population is now performing the most widespread experiment in drug interactions ever performed in the history of mankind!

There is an even more general reason to understand the principles of exchange of material into "compartments" in the human body. As Gallaher (1996) has indicated, the principles that underlie the flow of material into kinetic compartments are highly generic, and may be linked together to begin to produce what may be considered a system dynamic. For both of these reasons, it is essential to understand the concept that is known in pharmacokinetics as the Plateau Principle, and the kind of dynamic behavior that can be produced in a single container with one inflow and one outflow (often likened to a bathtub with a faucet and a drain, or to a swimming pool). Once one grasps this, it becomes easier to comprehend the effect of adding extra compartments to a model in such a manner that the inflow also has a characteristic rate of change, or there may be more than one route of elimination. Once that is understood, it is timely to consider whether the system truly behaves linearly, or if it might show interesting, nonlinear behavior.

18.1 Origins of the Plateau Principle

The simplest means of understanding the Plateau Principle is to consider a situation in which a substance that is not naturally present in the body is being introduced into the bloodstream at a constant rate. The initial concentration is zero, and the substance will accumulate until the rate at which it is eliminated equals the rate of infusion. During periods of constant infusion of drugs into the venous system, the concentration in plasma increases at a relatively constant rate, which can be represented by a zero-order rate constant (a pure number, k_i, with units of amount per minute or other time interval). However, the rate of elimination is often proportional to the amount of molecule present in the blood at any moment, and can be described by a first-order decay constant. This set of assumptions led to the development of the *Plateau Principle*, which states that compounds administered by various routes will approach a constant level at a rate determined by the elimination constant (Goldstein, et al., 1969; Pratt and Taylor, 1990; Gallaher, 1996). The same principle applies to turnover of isotopes incorporated into biological molecules; for several examples, refer to the recent review of Rudolf Schoenheimer's classic work, (Olson, 1997). Decay constants should govern the time course for change in any physical or chemical system in which the assumptions of zero-order filling and first-order decay are valid.

This kinetic model predicts that decay constants determine the period required for the concentration of enzymes to change, which was first recognized by Berlin and Schimke (1965) during comparative studies of enzymes in mammalian liver. They concluded that the period required for several enzymes to shift to new levels after hormonal stimulation was related to their half-lives, even when rates of synthesis were similar. The model suggests that the period required to achieve half of the total shift in concentration between two steady-state levels should equal the half-life of an mRNA or protein. Because the stability of these molecules varies greatly, so must the period required to achieve a new steady state vary.

18.2 The Plateau Principle Is Widely Applicable

It is true that the Plateau Principle was originally derived under rather restricted circumstances, but there are two reasons for its wide applicability. First, it is true of a large number of processes that elimination, loss, or degradation is proportional to the concentration of the substance of interest. Figure 18.1 illustrates how the same single-compartment model can be applied to water in a bathtub, a drug in the bloodstream, messenger RNA, or a receptor protein (Gallaher, 1996). Second, the Plateau Principle holds that the elimination rate determines the time course of shift between steady states regardless of how the shift is brought about. One must only modify this statement by adding that a substance that must pass through another compart-

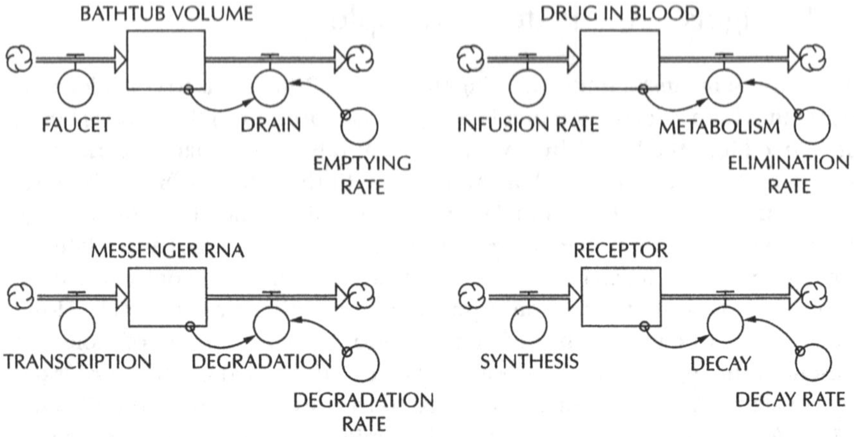

BATHTUB VOLUME

FAUCET   DRAIN

EMPTYING
RATE

DRUG IN BLOOD

INFUSION RATE   METABOLISM

ELIMINATION
RATE

MESSENGER RNA

TRANSCRIPTION   DEGRADATION

DEGRADATION
RATE

RECEPTOR

SYNTHESIS   DECAY

DECAY RATE

FIGURE 18.1

ment will experience a delay before the change in concentration begins, so that the accumulation curve will be sigmoidal. For example, nutrient absorption is delayed by the metering action of the stomach and by passage through the small intestine before the bloodstream is entered, so one expects a gradual rise, but it is really the same principle applied to three different sites.

What produces interesting behavior in the Plateau Principle is often the manner in which a substance enters the body. We usually consume food and drugs orally and periodically, and not by continuous infusion. Therefore, what must be done in a **STELLA**® model is to examine the premise that elimination constants govern the time course of the shift in concentration, and the way in which the timing of ingestion affects the outcome.

18.3 Outcomes Depend on Input Timing and Output Rates

Let us first ask how one can test the idea that the time course of accumulation of material, whether it be a drug, a nutrient, a messenger RNA, or a receptor protein, depends solely on its elimination rate. The easiest method is simply to increase the input rates to various levels using a fixed elimination rate, and plot the outcome as a fraction of the final steady-state value. Figure 18.2 shows the results if plotted in absolute (left side) or relative values (right side). When plotted as relative values, we can see that the time course is identical, and each of the curves completely overlaps.

The results just shown apply if the rate of infusion or synthesis is increased from one level to another and maintained at the new level until a new steady state is achieved. The Plateau Principle derived its name from the dependence of the time course on the elimination rate (or half-life) of the material in question. However, behavior of the model is more interesting if one imag-

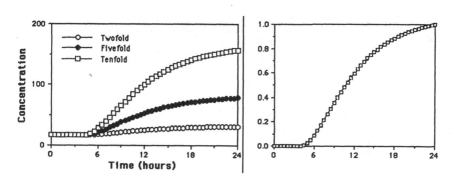

FIGURE 18.2

ines instances in which a specified amount is given at intervals, similarly to what is done in drug dosing or when one has meals throughout the day. Whether one is dealing with a nutrient or a drug, the goal is to achieve a desired effect. To do so, the amount of nutrient or drug in the blood must increase, and then it must accumulate in the tissues, so that one stops feeling hungry in the case of meals, or obtains enough vitamins to prevent deficiencies, or absorbs enough of a medicine to relieve fever or pain.

Basically, the reason we take in foods or drugs is to produce an effect. The effect is always concentration dependent, and it is possible to take in too much and produce a toxic effect. What is desired is to keep the amount of drug or nutrient within a certain concentration range, or therapeutic window. For instance, it is desirable to take in enough iron to prevent iron-deficiency anemia, but not enough to produce liver damage (hemosiderosis or hemochromatosis). To simulate this effect, we will now use the **STELLA®** model to introduce a substance in pulsatile fashion. The example shown in Figure 18.3 was generated by setting the elimination rate to 0.231 d^{-1} (half-life, 3 h), and the input rate was changed from pulses of 1 unit given every four hours to pulses of 4 or 8 units at the same intervals. The outcome is that the amount of substance in the blood rapidly increases after each pulse, and then decays at the value determined by the elimination rate. Under these conditions, the substance would achieve an equilibrium concentration of 4.3 units per volume if given continuously at the rate of 1 unit per interval, 8.6 units if given at 2 units, and 17.3 if given at 8 units per interval. If the aim was to maintain the concentration between values of 5–9, but never to allow it to equal 16, what would be the best choice of dosing intervals?

18.4 Elimination Rate, Daily Needs, and the Potential for Toxicity

Let us now use the simple model to examine the effect of the elimination rate on the accumulation of substances in the body. Except for glucose, most nutrients have much longer half-lives in the body than do most drugs. This is

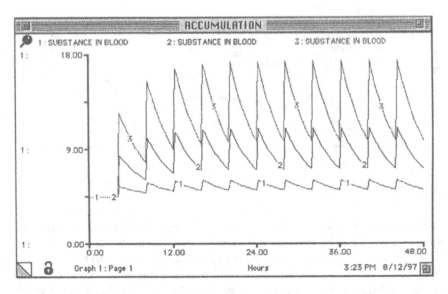

FIGURE 18.3

fortunate if the aim is to avoid deficiencies, but it leaves open the possibility of developing an excess or toxicity. Fortunately, there is a safety valve of sorts for water-soluble substances, because the kidney excretes compounds of low molecular weight, and only reabsorbs a portion of what is excreted. On the other hand, lipid-soluble substances are transported in the blood plasma bound to proteins or inside lipoprotein particles, so they are not readily excreted. For this reason, the body retains most of what is absorbed, and the concentration may build up to high levels in cell membranes or in storage sites within cells. It is notable that the two most toxic vitamins, vitamins A and D, are both fat soluble.

Consider the case of a water-soluble vitamin (vitamin C or ascorbic acid) and a fat-soluble vitamin (vitamin A or retinol). Table 18.1 shows the range of usual intakes in the U.S. population, the amount typically found in the body of a healthy adult, and the daily elimination rate expressed as a fractional rate (percentage eliminated per day). The RDA for vitamin C is 60 *milli*grams, and that for vitamin A is 1,000 *micro*grams (retinol equivalents), here converted to milligrams. Note that the amount in the body is similar, but the actual intake of vitamin C for men (around 100 mg per day) is a hundred times greater than the RDA value for vitamin A.

TABLE 18.1

	Usual intake (mg/day)	Amount in body (mg)	Elimination rate (per day)
Vitamin C	60–100	700–1,500	3.2% (0.032)
Vitamin A	1.0	315–877	0.5% (0.005)

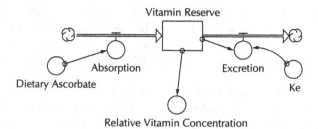

FIGURE 18.4

One could build a model using these absolute numbers, or compare the two nutrients in the relative sense by converting the values to multiples of the RDA. If one assumes an intake of one RDA for each vitamin, then the input can be 1.0 RDA, and the amount in the body could also be expressed in terms of multiples of the RDA. To determine initial steady-state values, divide the input of 1.0 by 0.032 for vitamin C and 0.005 for vitamin A. This is the value that should be used for the starting point in the model. Then we will increase and decrease the intakes and observe what happens. A diagram of the model is shown in Figure 18.4.

Equations for the Vitamin Model

```
Vitamin_Reserve(t) = Vitamin_Reserve(t - dt) +
(Absorption - Excretion) * dt
INIT Vitamin_Reserve = 31.25
INFLOWS:
Absorption = Dietary_Ascorbate * 1
OUTFLOWS:
Excretion = Vitamin_Reserve * Ke
Dietary_Ascorbate = 0.1
Ke = 0.032
Relative_Vitamin_Concentration = Vitamin_Reserve/0.3125
```

18.5 Simulate Effects of Changing the Intake of Vitamin C and Vitamin A

Let us compare results obtained first for changing our intake of vitamin C and then vitamin A. Open the Run menu and check that the default value for Time Specs is 365 and the unit of time is Days. It is reasonable to assume that deficiencies may develop within the course of a year; in some cases, it does not take this long. Select the graph tool or the table tool and place it on the page. Open it by double clicking. Select the function(s) you want to display from the list on the left and press the right arrow so the functions enter the Selected box. In this case, it is a good idea to check the box labeled Comparative so the outcomes will be saved each time. Click OK to return to the map.

Can the amount in the body be predicted? Yes: *The amount in the body equals the input rate divided by the output rate constant (or fraction used per day).* The input rate of (mg per day) divided by a rate constant (per day) gives mg. Run the model to ensure that you get a straight line. Click on the graph icon and highlight any item in the box on the right side. Now click the arrow on the right side of the item. Boxes that show minimum and maximum values will appear in the lower left. It is usually best to set the minimum to zero and the maximum to the maximum amount of material that is in the body. Then click on the Set button.

Now compare the following situations for vitamins C and A:

1. Set the input rate to 1.0 and the output rate to 0.032. Run the program for 365 days, and see what happens (with a graph or table open in the Comparative mode).
2. Change the input rate to 2.0 and compare the results; do the same with an input rate of 0, 0.1, and so on. How long does it take for a deficiency to develop? In the case of vitamin C, symptoms of scurvy often develop when the intake is 6–10 mg per day, which would be an input rate of 0.1 multiples of the RDA.
3. To compare outcomes for vitamin A, try the same thing with the output rate set to 0.005. How do the time courses compare? How long does it take to saturate the body in these cases, or to develop a deficiency? In the case of vitamin A, it may be necessary to select a longer time scale. Remember the rule that a steady state is achieved only after 5–6 half-lives have passed. Convert the fractional elimination rates to half-lives to determine what interval you should select for Time Specifications. Fig. 18.5 compares the results of decreasing the intakes of vitamins A and C to 0.1 multiples of the RDA (curves 1 and 2, respectively).
4. If toxicity were defined as twice the usual amount in the body, which case gets toxic first? And why it might be easier to develop a toxic effect from vitamin A or D than from vitamin C?

If one assumes that taking in the amount recommended by the RDA is sufficient to equal the amount used each day (plus a margin of safety), then it is interesting to compare the amount of nutrient stored in the body relative to the RDA. The numbers above suggest that well-nourished people maintain at least 12 RDAs worth of vitamin C in their tissue reserves, whereas they maintain over 300 RDAs of vitamin A. This makes sense in terms of the amount of time needed for a deficiency to occur. Strangely, many habitual diets are so poor that vitamin A deficiency is the most common worldwide nutritional deficiency, afflicting many millions of people with visual disorders,

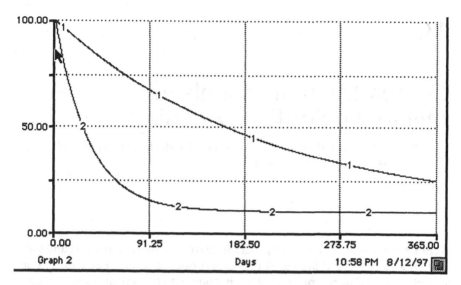

FIGURE 18.5

growth stunting, and blindness. It is difficult for a U.S. citizen to comprehend this, when a single carrot contains more than one RDA equivalent of beta-carotene.

References

Berlin, C.M., and R.T. Schimke. "Influence of turnover rates on responses of enzymes to cortisone." *Mol. Pharmacol.* 1 (1965): 149–156.

Gallaher, E.J. "Biological system dynamics: From personal discovery to universal application." *Simulation* 66 (1996): 243–257.

Goldstein, A., L. Aronow, and S. M. Kalman. *Principles of Drug Action.* New York: Harper and Row, 1969, 292–300.

Olson, Robert E. "The dynamic state of body constituents (Schoenheimer, 1939)," *J. Nutr.* 127 (1997): 1041S–1043S.

Pratt, W.B. and P. Taylor. *Principles of Drug Action, The Basis of Pharmacology*, 3rd ed. New York: Churchill Livingstone, 1990.

19

Compartmental Models in Metabolic Studies: Vitamin C

> It is . . . by model development and testing that the true power of modeling as a research tool reveals itself.
>
> David Foster and G. Hetenyi, Jr., 1991

STELLA® and similar computer programs allow one to use modeling and simulation as a tool for thinking. However, no one has time to squander; our modeling should have a purpose. Also, in coming to grips with questions, one must consider the rich source of data that is already available in the scientific literature. Modeling has been used as a quantitative tool to answer many significant medical questions, and simulation programs such as **STELLA®** may be used to interpret this information. Much of this information has been reported in a format derived from compartmental analysis. The purpose of this chapter is to introduce this method and ideas about using **STELLA®** to help interpret this kinetic information. Let us note that the comment quoted above concerning model development and testing implies a need to apply statistical methods to ensure that a model functions in a robust manner. However, it is reasonable for the student to use models that have already been developed with these research methods.

The availability of radioisotopes for biological research necessitated the development of a mathematical framework to understand kinetic data concerning metabolism. After the development of digital computing, this need was met in several ways. Notably, Dr. Mones Berman and colleagues generated the SAAM computer program and introduced a formal framework for analyzing data using compartmental models (Zech et al., 1986). This represents modeling conducted at the highest professional standard, and advanced students would be repaid manyfold by reading articles by Berman and colleagues (1982), Zech and colleagues (1986), Green and Green (1990), or Mazier and Jones (1994).

19.1 Compartmental Analysis

Recall that a kinetic compartment is a mathematical construct that may or may not correspond to material in a physical space. The liver, for instance, contains cholesterol in storage forms and in forms being used for synthesis of bile acids and low-density lipoprotein particles. These behave in distinct ways, so it may be necessary to include several compartments for cholesterol in this organ. Similarly, cholesterol is found in several distinct particles in the blood, which again cannot be treated as one unit although they coexist in the blood plasma. The single kinetic compartment whose behavior is governed by the Plateau Principle is the element on which compartmental analysis is based. In the stock-and-flow diagram of **STELLA®**, a stock (rectangle) represents a single compartment. It is easy to imagine the behavior of systems in which several compartments or stocks are linked together, with each one receiving input from another and providing outflows into other compartments. Each compartment responds to an input with an asymptotic approach to a new level, and decays exponentially; this behavior is identical to a mathematical function written as $C * e^{-kt}$. By linking several stocks and flows in a **STELLA®** diagram, one obtains the same results that a mathematician obtains by adding several exponential terms together. This is useful because one may use **STELLA®** to interpret data that has been obtained in hundreds of scientific investigations in which a technique called *compartmental analysis* was used to study metabolism of drugs, nutrients, and substances in the body.

The need for compartmental analysis in biology arose from the use of tracers to study metabolism; for instance, glucose, an amino acid, or a fatty acid containing an isotope of carbon, hydrogen, or nitrogen would be used (Green and Green, 1990; Mazier and Jones, 1994). In many cases, analysis involved infusion of radioactive compounds into animals or human subjects. The studies by Baker and colleagues (1969) with ascorbic acid are a classic example of this method. Samples were generally taken from the blood and urine at intervals after the injection, and attempts were made to extract as much information as possible from the change in amounts of radioactivity with time. The analysis assumed that the radioactive form of the compound (the tracer) behaved identically to the compound being traced (the tracee), and that mixing was complete and rapid. In many cases, especially with human subjects, it is not possible to take other tissue samples, so it is important to obtain as much information as possible from the accumulation and decay of the tracer in the blood plasma.

The conventions used by most professional kineticists are shown in Figure 19.1 (adapted from Green and Green, 1990). A compartment or "stock" is usually represented as a circle. A flux of material into the system is designated with a U, and material in a compartment is indicated by M for mass followed by a number designating the compartment in which it is found. Most, but not all, kineticists indicate the fluxes or rates of transfer between compartments by a letter R (I, J) to indicate flux from compartment J to compartment I. Thus, R(1,2) indicates movement *from* 2 to 1 in the figure. This flux is usually equal to the product of a fractional transfer coefficient L(1,2) multiplied by M(2). Presumably the L designation originated in the convention that the Greek letter, λ, designated first-order rate constants. Finally, efflux from the system is designated here by the rate R(0,2), indicating that material is moving out of the system and into the compartment of oblivion — equivalent to the cloud symbol in a **STELLA®** diagram.

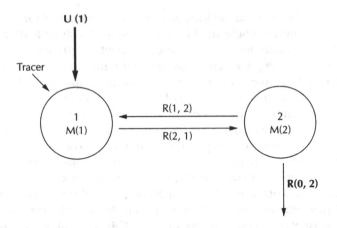

FIGURE 19.1

Not all compartmental analysis requires radioactive tracers. For instance, if one can measure the concentration of a drug in the blood and urine, it can be tracked directly because no drug is present naturally in the body. In the case of studies of nutrient metabolism, much information can be obtained by eliminating the nutrient from the diet, following its loss from the blood over time, and then providing it again at various levels. For these reasons, an immense amount of information concerning the metabolism of drugs and nutrients is available in the form of compartmental models. The information in these models is only accessible through the use of computers to solve the equations.

19.2 A Note About Linearity

The Plateau Principle describes the response of a linear system. In mathematics, a linear system is one in which the solution to an equation only involves one or more constants multiplied by a variable; for instance, an elimination rate may equal $-k * C$, where k is a constant and C is a concentration. Strictly speaking, the equations that describe flux of material through a compartmental system are linear because the inputs and outputs can be treated as the product of constants multiplied by single variables. On the other hand, nonlinear systems are those in which the solutions require that variables be multiplied in some combination, for instance, $C * E$, in which E may be a concentration of an enzyme or another substrate. Consider that nutrient uptake through the intestinal wall depends upon the concentration of transporter molecules in the absorptive cell, and in most cases requires the cotransport of sodium, which then must be eliminated from the cell by exchange with potassium ion in an ATP-dependent manner, and it is evident that almost any system of interest to biologists is nonlinear. One might say, facetiously, that a linear biological system is one that has not been fully investigated!

19.3 Compartmental Analysis of Nutrient Metabolism

By far the most widespread use of compartmental analysis is in the areas of pharmacokinetics and pharmacodynamics. Clinical pharmacologists must monitor levels of drugs used for cancer chemotherapy, for instance, and ensure that levels in the blood plasma maximize effectiveness but minimize toxicity. Gratifyingly, it is possible to do this using simple, two- or three-compartment models. However, similar methods are crucial to anyone interested in the use of metabolic fuels, vitamins, and mineral elements, or understanding important problems that range from lipoprotein metabolism, fat metabolism, protein synthesis and turnover, to wasting diseases in which tissues are broken down in diseases or after trauma.

A recent conference on experimental nutrition summarized examples of compartmental analysis of several vitamins and minerals (Coburn and Townsend, 1996). Let us take one example from the literature and illustrate the use of **STELLA**® and similar programs to analyze the metabolism of vitamin C.

19.4 A Multicompartment Model of Ascorbic Acid Metabolism

Humans, guinea pigs, and a few other animals lost the ability to synthesize ascorbic acid (vitamin C) from glucose when our ancestors acquired a mutation in a gene needed for its biosynthesis (Gerster, 1987). Molecular cloning has demonstrated that the gene encoding L-gulonolactone oxidase contains such extensive deletions and rearrangements that it no longer functions to produce messenger RNA, and is a pseudogene (Nishikimi et al., 1994). For this reason, ascorbic acid is now nutritionally essential, and the research community continues to seek functional markers that help define intakes that may be optimal (Levine et al., 1991). This requirement produces a question concerning optimal dosing regimens: How much and how often should one consume vitamin C? This question is ideally suited for kinetic modeling, so let us examine the literature concerning models of vitamin C metabolism.

Kallner and colleagures (1979) developed a three-compartment model of ascorbic acid kinetics in humans to assist in analyzing studies of total amounts and turnover of vitamin C using $(1-{}^{14}C)$-ascorbic acid as a tracer. A diagram indicating the kinetic constants they reported is shown in Figure 19.2; it is a simple matter to reproduce this model using **STELLA**® or other simulation software. The three compartments will be assumed (hypothetically) to be (1) the blood, (2) the tissues (a rapidly exchanging compartment), and (3) the extracellular fluid (ECF; a slowly exchanging compartment). Concerning the half-life, note that there are two important elimination constants. $K(0,1)$ is elimination in the urine and $K(0,2)$ is metabolic destruction. Although vitamin C can be partially regenerated after oxidation, it is converted to oxalic acid and eliminated in the urine. Just as with glucose, there is a

Model for Human Ascorbic Acid Kinetics

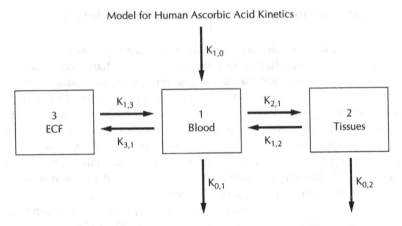

FIGURE 19.2

threshold concentration in the blood beyond which the vitamin cannot be reabsorbed. For this reason, there is no fixed elimination rate; rather, it depends on dosage and other factors such as whether a subject smokes tobacco or has diabetes mellitus, so that the half-life can vary from 8–40 days. Similarly, absorption depends on dosage, which can range from about 93% at 30 mg/d to as little as 12% at intakes of 500-1000 mg/d (Kallner et al., 1979; Piotrovskij et al., 1993).

Table 19.1 lists comparative data for vitamins C and A. The data for vitamin C will be used to generate a multicompartment model of vitamin C metabolism that could be used to ask the effects of taking different doses of vitamin C. If one believes that the RDA is too low and that benefit would be accrued by a higher intake, then how much should one take and at what interval? Commercially available supplements vary from 30 mg to 1000 mg, in addition to the high level found in citrus fruits and other dietary sources, and

TABLE 19.1 Vitamin C compared to vitamin A

	Vitamin C	Vitamin A
Molecular weight	176	537
Adult RDA	60 mg	800–1000 μg (retinol equiv.)
Uptake	Na$^+$-dept. transport (90% absorbed)	Absorbed with lipid and esterified
Storage sites	All tissues; adrenal	Liver (90%)
Tissue concentration	ca. 10 mg/100 g	0.15 mg/g liver
Total body pool	1000–3000 mg	315–877 mg
Blood plasma	0.4–1.0 mg/dL	30–40 μg/dL
Utilization rates	ca. 40 mg/day	ca. 900 μg/day
Biological half-life	7–42 d (Avg. = 12) (3% d^{-1})	128–156 days (0.5% d^{-1})
Excretion	Renal, as oxalate or ascorbate	Fecal, conjugated Renal (oxidized)
Toxicities	Over 1 gram/day (Questionable)	Over 15 mg/day (Acute and chronic)

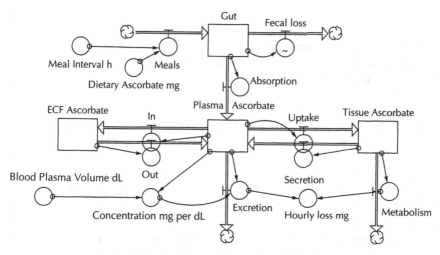

FIGURE 19.3

many people do not realize that the RDA level is two standard deviations above the average requirement!

The three-compartment model of Kallner and colleagues (1979) includes four exponential terms, with the extra one identified as the input from the gastrointestinal tract. The model was translated into a **STELLA**® diagram with minor modifications. All units were left as originally reported, with units of mg of vitamin per deciliter of blood, and rates in hourly units. The changes were as follows: (1) input variables that allow the user to control the amount of ascorbate per meal and the meal interval from the high level map; (2) a term that reduces absorption due to increased fecal loss when large supplements are taken; and (3) addition of an elimination rate that acts as a threshold for reabsorption in the kidney. The threshold is near 0.7 mg per dL; in this case, it was set to 0.9. A small amount of unmodified ascorbic acid is always lost in the urine, but the term used here causes the vitamin to be cleared with a tenfold higher fractional rate when the concentration in the blood plasma exceeds the threshold. Figure 19.3 shows the **STELLA**® model.

Equations for the Ascorbate Model

```
ECF_Ascorbate(t) = ECF_Ascorbate(t - dt) + (In - Out)
* dt
INIT ECF_Ascorbate = 241
INFLOWS:
In = Plasma_Ascorbate * 0.057
```

```
OUTFLOWS:
Out = ECF_Ascorbate * 0.045
Gut(t) = Gut(t - dt) + (Meals - Absorption -
Fecal_loss) * dt
INIT Gut = 30
INFLOWS:
Meals = PULSE(Dietary_Ascorbate_mg,8,Meal_Interval_h)
OUTFLOWS:
Absorption = Gut * 2.2
Fecal_loss = GRAPH(Gut)
(0.00, 0.00), (100, 10.0), (200, 20.5), (300, 29.5),
(400, 39.5), (500, 48.5), (600, 58.5), (700, 67.5),
(800, 76.0), (900, 84.0), (1000, 90.0)
Plasma_Ascorbate(t) = Plasma_Ascorbate(t - dt) +
(Secretion + Absorption + Out - Excretion - Uptake -
In) * dt
INIT Plasma_Ascorbate = 192
INFLOWS:
Secretion = Tissue_Ascorbate * 0.186
Absorption = Gut * 2.2
Out = ECF_Ascorbate * 0.045
OUTFLOWS:
Excretion = IF(Concentration_mg_per_dL< = 0.9)
THEN(Plasma_Ascorbate * 0.0021)
ELSE(Plasma_Ascorbate * 0.231)
Uptake = Plasma_Ascorbate * 0.693
In = Plasma_Ascorbate * 0.057
Tissue_Ascorbate(t) = Tissue_Ascorbate(t - dt) +
(Uptake - Secretion - Metabolism) * dt
INIT Tissue_Ascorbate = 717
INFLOWS:
Uptake = Plasma_Ascorbate * 0.693
OUTFLOWS:
Secretion = Tissue_Ascorbate * 0.186
Metabolism = Tissue_Ascorbate * 0.0029
Blood_Plasma_Volume_dL = 34.25
Concentration_mg_per_dL = Plasma_Ascorbate *
0.143/(Blood_Plasma_Volume_dL)
Dietary_Ascorbate_mg = 30
Hourly_loss_mg = Excretion + Metabolism
Meal_Interval_h = 12
Total_Ascorbate_mg = ECF_Ascorbate + Plasma_Ascorbate +
Tissue_Ascorbate
```

19.5 Volume of Distribution

The values entered for each compartment are total amounts, originally reported in terms of mg per kg body weight. Based on these quantities, the amount in the accessible compartment is 192 mg. This compartment is labeled Blood Plasma in the diagram, but it is evident that kinetic compartment 1 is not equivalent to the blood plasma alone, because the usual concentration of ascorbic acid in the blood is about 0.4–0.8 mg per dL, so that 3 L of plasma should only contain about 12–24 mg of ascorbate. The difference is explained by the concept of *volume of distribution*, which is the "apparent" volume of kinetic compartment 1, or the factor by which the concentration in blood must be multiplied to obtain the total amount in a tissue. Volumes of distribution are defined as the amount of space a substance would take up if the concentration was the same in the original compartment and the destination. The apparent space is often much larger than equivalent physical spaces because active transport concentrates nutrients.

Volume of distribution =
Amount in kinetic compartment divided by plasma concentration

or

Plasma concentration =
Amount in kinetic compartment/Volume of distribution

If tissues did not concentrate the substance, the total volume of distribution would equal the volume of fluid in the patient's body, about 42 liters. However, tissues do concentrate the vitamin, so it is not unusual for the volume of distribution to be several times larger than the volume of the body.

19.6 Clearance

Just as the idea of volume of distribution needed to be developed to relate the concentration of a substance in the blood to the concentration in the tissues or kinetic compartments, an idea is needed to relate concentration to its rate of elimination from the body. That relationship is called clearance, CL:

Rate of elimination = Concentration $\times$ Clearance = C $\times$ CL

The *units* of clearance are similar to flow, *volume/unit time*. Example: If clearance is 1 liter/hour, and the concentration is 1 mg/liter, what is the rate of elimination? (1 mg/hour.) Remember that first-order rate constants always have units of $time^{-1}$. This means that clearance has two parts, a rate constant and a volume,

CL = kV, $(time^{-1})$(volume), for instance, liters per hour.

So the rate of elimination also equals kVC, with units of quantity per time. Happily, we can now relate two obscure concepts to one another, for clearance is related to the volume of distribution:

$$V = CL/k$$

19.7 Routes of Elimination

Vitamins can be eliminated in three ways:

1. By metabolic use, such as oxidation of vitamin C or E
2. By excretion from the kidneys
3. By excretion into the stool

For vitamin C, only the first two routes are used; fecal vitamin C is simply not absorbed, so it never enters the kinetic compartment. As intake increases, vitamin C absorption decreases from about 92% to as little as 20%. On the other hand, the major route of excretion for the fat-soluble vitamin A is into the bile and stool.

19.8 Predictions Based on the Model of Vitamin C Metabolism

The only valid reason to supplement dietary intake of a nutrient must be based on the idea that the vitamin's functions or protective effects are related to tissue concentrations of the nutrient. Tissue effects are almost always saturable processes, which can be represented by Michaelis-Menten type kinetics. Based on this concept, Levine and colleagues. (1991) demonstrated that the rate of norepinephrine synthesis was maximal at a concentration of 2-10 mmol/L in the secretory granules that produce norepinephrine, and biosynthesis appeared to be saturated at usual plasma levels due to the active uptake process. Bearing this in mind, let us ask how the model predicts that plasma ascorbic acid varies as a function of dietary or supplemental intake.

The model may be used to follow the course of depletion or repletion as dietary intakes vary from 0 to 500 mg per meal. Because the half-life of the vitamin is about 2 weeks, extensive depletion would require one to use an extended time frame. Instead, let us examine what occurs over a few days of supplementation.

Curve 1 on the comparative graph (Fig. 19.4) is the usual situation with the subject taking in 60 mg per day in two equal amounts, delivered as pulses. Each pulse can be seen as a small, temporary increase in the concentration curve. Curve 2 is the response predicted after taking in two supplements of 250 mg of vitamin C. The response is that the augmented amount causes a brief pulse that increases the plasma concentration modestly, but the average

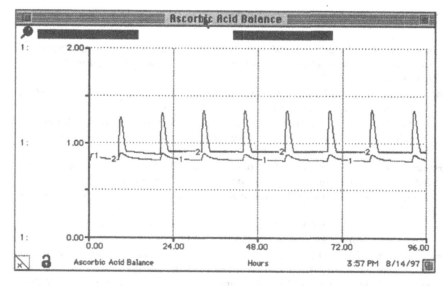

FIGURE 19.4

concentration barely increases. The advocate might hold that these brief increases are beneficial, but that argument should be weighed relative to the knowledge that any antioxidant can be a pro-oxidant. A standard method of producing free-radical damage is to add ascorbic acid to a minute amount of dissolved iron or copper!

The second question to ask is: What happens to the total tissue reserves of the vitamin. The inital amount in the subject's body was about 1160–1180 mg. After increasing the intake from 60 mg per day to 500 mg per day, the new values ranged between 1300–1400 mg. An 8-fold increase in intake produced a 0.15-fold (15%) increase in tissue stores! This is true; it is possible for a well-nourished person to increase tissue stores from about 1500 mg to as much as 2000 mg by heavy supplementation, and for most people, no evident adverse effects occur when ascorbate is ingested at levels of 1–2 g per day. Some proponents have argued that the amount that is naturally synthesized in the bodies of animals (for instance, rats and pigs) that can still synthesize ascorbate is as much as ten times higher than the RDA if expressed per kg of body weight (Levine, 1986). Still, this would suggest that no one needs more than 0.5 g per day, preferably taken in small amounts that would mimic the natural biosynthetic process.

19.9 Challenge to the Student

Consider devising a model of vitamin A metabolism to compare the outcome of supplementation to what is observed with vitamin C.

References

Baker, E.M., R.E. Hodges, J. Hood, H.E. Sauberlich, and S.C. March. "Metabolism of ascorbic-1-^{14}C-acid in experimental human scurvy." *Am. J. Clin. Nutr.* 22 (1969): 549-558.

Berman, M., S.M. Grundy, and B.V. Howard, eds. *Lipoprotein Kinetics and Modeling*. New York: Academic Press, 1982.

Coburn, S.P., and D.W. Townsend. "Mathematical modeling in experimental nutrition: Vitamins, proteins, methods." *Adv. Food Nutr. Res.* 40, San Diego: Academic Press, 1996.

Foster, D.M., and G. Hetnyei, Jr. "Role of modeling in the design of experiments in carbohydrate metabolism." *JPEN-J. Parenter. Enteral Nutr.* 15 (1991): 67S-71S.

Gerster, H. "Human vitamin C requirements." *Ztsch. Ernahrungswissenschaft* 26 (1987): 125-137.

Green, M.H., and J.B. Green. "The application of compartmental analysis to research in nutrition." *Annu. Rev. Nutr.* 10 (1990): 41-61.

Kallner, A., D. Hartmann, and D. Hornig. "Steady state turnover and body pool of ascorbic acid in man." *Am. J. Clin. Nutr.* 32 (1979): 530-39.

Levine, M. "New concepts in the biology and biochemistry of ascorbic acid." *N. Engl. J. Med.* 314 (1986): 892-902.

Levine, M., K.R. Dhariwal, P. W. Washko, J. DeB. Butler, R.W. Welch, Y. Wang, and P. Bergsten. "Ascorbic acid and in situ kinetics: a new approach to vitamin requirements." *Am. J. Clin. Nutr.* 54 (1991) 1157S-1162S.

Mazier, M.J.P., and P.J.H. Jones, "Model-based compartmental analyses in nutrition research." *Can. J. Physiol. Pharmacol.* 72 (1994): 415-422.

Nishikimi, M., R. Fukuyama, S. Minoshima, N. Shimizu, and K. Yagi. "Cloning and chromosomal mapping of human nonfunctional gene for L-gulono-gamma-lactone oxidase, the enzyme for L-ascorbic acid biosynthesis missing in man." *J. Biol. Chem.* 269 (1994): 13685-13688.

Piotrovskij, V.K., Z. Kallay, M. Gajdos, M. Gerykova, and T. Trnovec. "The use of a nonlinear absorption model in the study of ascorbic acid bioavailability in man." *Biopharmaceutics Drug Disp.* 14 (1993): 429-442.

Zech, L.A., R.C. Boston, and D.M. Foster, "The methodology of compartmental modeling as applied to the investigation of lipoprotein metabolism." *Meth. Enzymol.* 129 (1986): 366-383.

Zech, L.A., D.J. Rader, and P.C. Greif, "Berman's simulation analysis and modeling." *Adv. Exp. Med. Biol.* 285 (1991): 188-99.

20

Circadian Rhythms

Now that we know the viscera can be taught, the thought comes that we've been neglecting them all these years.

Lewis Thomas,
The Lives of a Cell, 1974, p. 66

There is nothing so commonplace as a day, or so extraordinary, for the period of time that we customarily divide into 24 hours depends upon the amount of time required for our particular planet to spin one full turn upon its axis. Move to the moon, Mars, or Jupiter, and the day takes on quite different dimensions (if by a "day" we mean one complete cycle of light and darkness). If the earth rotated slower, the day would be longer; if it did not rotate at all, we would either be bathed in perpetual sunlight, or immersed in constant darkness. Needless to say, if absolute time exists, it has nothing to do with day length!

Many of our rhythms are keyed to the cycle of light and darkness for the very good reason that we must eat, and most creatures have synchronized their feeding activities according to the level of daylight. Many circadian and circannual rhythms are coordinated by the level of light that passes from the eye to the superior cervical ganglion and then to the pineal organ, the source of melatonin. Nonetheless, it is well that many rhythms not be modified by the time of day, for the pacemakers of the heart and the respiration center must continue at all times if we are to survive. With just a moment's consideration (however a moment is defined), it is evident that celestial time is relative, and that biological time includes some rhythms keyed to external cues of light, temperature, and food availability, whereas others are independent. If one begins to list all the different periodic aspects of our physiology (body temperature, hunger, micturition, hormone secretion, reproduction . . .), it soon becomes evident that our systems contain multiple oscillators and that many of these oscillators depend upon one another.

In the book, *From Clocks to Chaos*, Glass and Mackey (1988) mention six kinds of rhythmic or oscillatory systems: pacemakers, central pattern generators, mutual inhibitions, sequential disinhibition, negative feedback systems, and mixed feedback systems with time delays. Models of all these oscillations may be constructed in various ways. Let us consider two kinds of oscillatory systems.

20.1 The van der Pol Oscillator

Our sense of time and our clocks depend upon oscillators. Many early clocks used pendulums and escapements to mark time, and now electronics are used to mark intervals. Mathematically, oscillations occur when two or more time-dependent variables interact. One of the earliest models of rhythmic behavior, the van der Pol oscillator, was first described in reference to an electrical circuit in which current and voltage fluctuated over time (Van der Pol, 1934). The equations were modified for the purpose of describing human circadian functions by Wever (1965, 1984) and Kronauer and colleagues (1982). A simple model of two interdependent harmonic functions is shown in Figure 20.1 along with representative results in Figure 20.2.

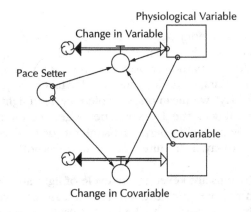

FIGURE 20.1

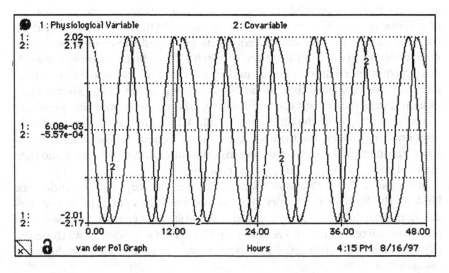

FIGURE 20.2

Equations for a van der Pol Oscillator

```
Covariable(t) = Covariable(t - dt) +
(Change_in_Covariable) * dt
INIT Covariable = 1
INFLOWS:
Change_in_Covariable = -Pace_Setter *
Physiological_Variable
Physiological_Variable(t) = Physiological_Variable(t -
dt) + (Change_in_Variable) * dt
INIT Physiological_Variable = 2
INFLOWS:
Change_in_Variable = 1/Pace_Setter * (Covariable-
((Physiological_Variable^3)/3) +
Physiological_Variable)
Pace_Setter = 0.5
```

Try changing the value of the pacesetting input to learn how the oscillations change. The frequency decreases as the number is made smaller. Circadian rhythms are almost never perfectly regular, and may contain one or more peaks during a 24-h period. For instance, the circadian rhythm in cortisol secretion (or corticosterone in the rat) typically has a single peak, whereas rhythms keyed to feeding are related to the onset of each meal. Wever (1984) modified the oscillator so that it would only take on positive values and its stability would be improved. The equations are particularly interesting because it is possible to link them to an external forcing function; for instance, Klerman and colleagues (1996) used this system to simulate the effects of light on the human circadian pacemaker. This set of equations is very useful for contemplating the nature of intrinsic biological rhythms that may be reset by a time-keeping external stimulus.

One drawback to a function such as the van der Pol oscillator is that it is not clear what connection the model variables may have to physiological functions. What does it mean to take the second derivative of cortisol secretion, for instance? Although the van der Pol oscillator is an important example of behavior of a periodic system, it is worth evaluating other means of achieving rhythmic behavior.

20.2 A Generic Oscillator

Oscillations occur in many circumstances under which one variable affects the rate of production of a second variable, and the second one in turn modifies the production or elimination of the first. A simple model in which

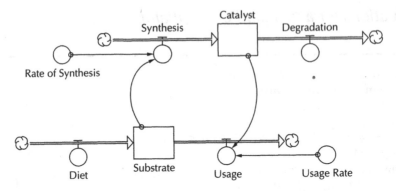

FIGURE 20.3

an item regarded as a catalyst affects the level of a substrate (here obtained from the diet) that is destroyed by the catalyst is shown in Figure 20.3. Try changing the initial quantities of these two variables and modifying their rates of production and destruction, and learn what happens to the rhythm. Under certain conditions, the oscillations cannot be maintained and will approach a limiting value.

Equations for the Catalyst Oscillator

```
Catalyst(t) = Catalyst(t - dt) + (Synthesis -
Degradation) * dt
INIT Catalyst = 1
INFLOWS:
Synthesis = Substrate * Rate_of_Synthesis
OUTFLOWS:
Degradation = 1
Substrate(t) = Substrate(t - dt) + (Diet - Usage) * dt
INIT Substrate = 15
INFLOWS:
Diet = 10
OUTFLOWS:
Usage = Catalyst * Usage_Rate * 10
Rate_of_Synthesis = 0.1
Usage_Rate = 1
```

20.3 Circadian Rhythms in Synthesis of Specific Proteins

It goes without saying that time delays occur to some extent in all systems, and that this contributes to patterning of behavior. Let us go to the research literature for information that can be used to generate a model in which negative feedback occurs and observe effects on daily rhythms. This is a situation in which the observed rhythms depend on interactions among other underlying cues or pacemakers, such as onset of light or darkness, meal timing, or hormonal rhythms (Boulos and Terman, 1980).

Enzyme induction generally refers to an increase in enzyme activity irrespective of mechanism. The change in activity can be due to activation of preexisting enzyme, increased biosynthesis, or decreased degradation. One of the earliest instances of this was a daily rhythm observed in the activity of tyrosine aminotransferase and other (but not all) enzymes that are involved in amino acid metabolism in rat liver (Watanabe et al., 1968; Fernstrom et al., 1979). Basically, the enzyme activity increases after the animals begin to feed (rats are nocturnal), and the magnitude of the change depends on the amount of protein in the diet. With typical rodent diets, the change may only be a few fold, but it is possible to obtain changes of 10–20-fold with certain schedules of feeding high protein diets. Furthermore, the activity depends upon a variety of other factors, including secretion of the hormones, insulin and glucagon, a rhythm in the adrenal steroid, corticosterone, and activity of nerves or other factors from the brain. The model shown in Figure 20.4 was generated to qualitatively reproduce the effect of meal timing and protein content on amino acids and an enzyme that metabolizes an amino acid (tyrosine, a precursor for hormones and neurotransmitters).

The high level controls allow the dietary protein content (stated as a fraction of the total diet) to be varied from 0 to 0.5 (Fig. 20.5). In addition, the half-lives of the mRNA and protein that are induced by the amino acid supply

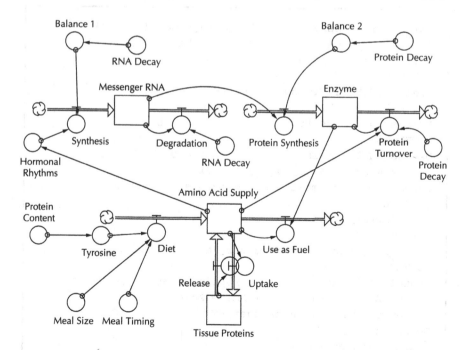

FIGURE 20.4

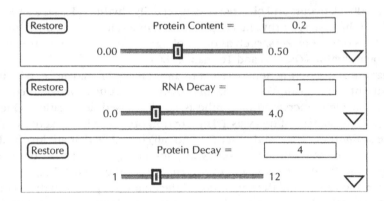

FIGURE 20.5

and hormonal balance may be varied; for the mRNA, the range is 0-4 h and for the protein, it is 1-12 h. Most mRNAs have shorter half-lives than the proteins they encode. These controls will allow the user to test the effect of stabilizing or destabilizing an mRNA or protein has on the ability to control its level.

Equations for the Circadian Oscillator

```
Amino_Acid_Supply(t) = Amino_Acid_Supply(t - dt) +
(Diet + Release - Use_as_Fuel - Uptake) * dt
INIT Amino_Acid_Supply = 0.2
INFLOWS:
Diet = IF(Meal_Timing> = 18) THEN(Meal_Size * Tyrosine
* 0.693) ELSE(Meal_Size * Tyrosine * 0.05)
Release = Tissue_Proteins * 2.6 * 10^-6
OUTFLOWS:
Use_as_Fuel = Amino_Acid_Supply * 0.08 * Enzyme
Uptake = Amino_Acid_Supply * 0.01
Enzyme(t) = Enzyme(t - dt) + (Protein_Synthesis -
Protein_Turnover) * dt
INIT Enzyme = 1.4
INFLOWS:
Protein_Synthesis = Messenger_RNA * 2 * Balance_2
OUTFLOWS:
Protein_Turnover = (0.693/Protein_Decay) * Enzyme *
(Amino_Acid_Supply/(0.2 + Amino_Acid_Supply))
Messenger_RNA(t) = Messenger_RNA(t - dt) + (Synthesis -
Degradation) * dt
INIT Messenger_RNA = 0.1
```

```
INFLOWS:
Synthesis = Hormonal_Rhythms * 0.1 * Balance_1
OUTFLOWS:
Degradation = Messenger_RNA * (0.693/RNA_Decay)
Tissue_Proteins(t) = Tissue_Proteins(t - dt) + (Uptake
- Release) * dt
INIT Tissue_Proteins = 1500
INFLOWS:
Uptake = Amino_Acid_Supply * 0.01
OUTFLOWS:
Release = Tissue_Proteins * 2.6 * 10^-6
Balance_1 = 1/RNA_Decay
Balance_2 = 2/Protein_Decay
Hormonal_Rhythms = 3 * (Amino_Acid_Supply^2)/(0.4^2 +
Amino_Acid_Supply^2)
Meal_Size = 20
Meal_Timing = MOD(TIME,24)
Protein_Content = 0.4
Protein_Decay = 2
RNA_Decay = 1
Tyrosine = 0.05 * Protein_Content
```

A typical pattern of regulation is shown in Figure 20.6. Observe that under the conditions shown, rats are nocturnal creatures that begin eating when the lights go off at around 6 PM. A rise in plasma amino acids occurs shortly

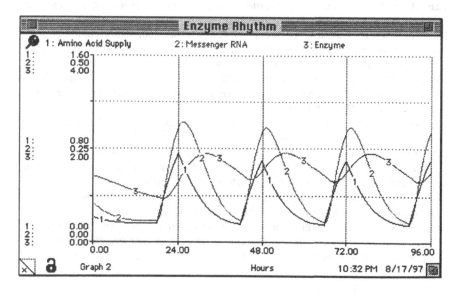

FIGURE 20.6

thereafter, and this affects hormones that increase the rate of synthesis of the messenger RNA for enzymes involved in amino acid metabolism. Note that the peak of the mRNA accumulation occurs after the amino acid supply has already started to decline. The increased mRNA gives rise to increased enzyme synthesis, which does not peak until after the amino acids in the blood plasma have declined substantially. This combination of negative feedback and time delay gives rise to oscillations that continue because rats normally eat each day, thereby producing a new stimulus for mRNA synthesis.

Try changing the protein content of the diet, the mRNA half-life, and the protein half-life in various combinations and see what happens. You will begin to get the idea that it is not a trivial matter to couple the rate of usage of a nutrient in the bloodstream with its availability, especially when the catalyst for its usage depends on a precursor (the mRNA). This fine-tuning has to be accomplished for twenty different amino acids with interactions among all the organs in the body and many distinct sets of genes (including amino acid transporters and catabolic enzymes). Our ability to survive a fast would be severely hindered without the ability to release amino acids from muscle and convert them to glucose and ketone bodies at appropriate rates.

References

Boulos, Z., and M. Terman. "Food availability and biological rhythms." *Neurosci. Biobehav. Rev.* 4 (1980): 119-131.

Fernstrom, J., R.J. Wurtman, B. Hammarstrom-Wiklund, W.M. Rand, H.N. Munro, and C.S. Davidson. "Diurnal variations in plasma concentrations of tryptophan, tyrosine, and other neutral amino acids; effect of dietary protein intake." *Am. J. Clin. Nutr.* 32 (1979): 1912-1922.

Glass, L., and M.C. Mackey. *The Rhythms of Life. From Clocks to Chaos*. Princeton, N.J.: Princeton University Press, 1988.

Klerman, E.B., D.-J. Dijk, R.E. Kronauer, and C.A. Czeisler. "Simulations of light effects on the human circadian pacemaker: implications for assessment of intrinsic period." *Am. J. Physiol.* 270 (1996): R271-R282.

Kronauer, R.E., C.A. Czeisler, S.F. Pilato, M.C. Moore-Ede, and E.D. Witzman. "Mathematical model of the human circadian system with two interacting oscillators." *Am. J. Physiol.* 242 (1982): R3-R17.

Van der Pol, B. "The nonlinear theory of electric oscillations." *Proc. I.R.E.* 22 (1934): 1054-1086.

Watanabe, M., V.R. Potter, and H.C. Pitot. "Systematic oscillations in tyrosine transaminase and other metabolic functions in liver of normal and adrenalectomized rats on controlled feeding schedules." *J. Nutr.* 95 (1968): 207-222.

Wever, R.A. "A mathematical model of circadian rhythms." In: *Circadian Clocks*. J. Aschoff, ed., Amsterdam: North Holland, 1965, 47-63.

Wever, R.A. "Toward a mathematical model of circadian rhythmicity." In: Moore-Ede, M.C., and C.A. Czeisler, Eds., *Mathematical Models of the Circadian Sleep-Wake Cycle*. New York: Raven Press, 1984, 17-79.

21

Diet Composition and Fat Balance

And the Lord spoke unto Moses, saying, Speak unto the children of
Israel, saying, *Ye shall eat no manner of fat . . .*

Leviticus 7:22–24

Hearken diligently unto me, and eat ye that which is good, and *let your
soul delight itself in fatness.*

Isaiah 55:2, The King James Bible

21.1 Introduction

People who have a difficult time maintaining a desirable body weight are experiencing the First Law of Thermodynamics with a vengeance. Once it is
lodged in adipose tissue, body fat persists. To remove triacylglycerol from
the fat cell, fatty acids must be mobilized by hormonally activated lipolysis
and oxidized in mitochondria in muscles and other tissues. This trait perfectly suits our only truly long-term energy reserve. And adipose tissue is a
marvelous invention; it stores about 4,000 kilocalories (17 MJ) per pound of
pure fat, or 8,800 kilocalories (37 MJ) per kilogram of pure triacylglycerol.
It insulates us against loss of heat, cushions our internal organs, and provides metabolic water when it is oxidized, so that it is equally useful in the
chill of winter or the parched heat of late summer. Yet this marvelous substance has been discredited by fashion designers who have a sense of style
but none of evolution and its mysteries. Alas, our labor-saving devices have
now conserved so much of our energy, and our restaurants make so much
available so readily, that many millions of people have become obese; non-
insulin-dependent diabetes, once rare, has become commonplace, and death
comes to the unmindful by what is called the Metabolic Cardiovascular Syndrome, with blood clots blocking one or more coronary arteries.

STELLA® can help one learn about the accumulation and daily use of body
fat. It can demonstrate the quantitative relationship between energy consumed and energy expended. Yet one should be warned, if the concept of fat
balance is exceedingly simple, in practice, there are certain complexities.
The complexities come about because our bodies cannot leave energy metabolism to chance; lack of energy means failure to thrive, and often, failure
to survive. Thus it is an oversimplification to say that fat balance is just a matter of the difference between what is consumed and stored, versus what is
mobilized and oxidized. This is true, but it shall turn out that one must also
consider the body's ability to regulate the rate at which all energy-yielding nutrients are oxidized, and to alter the proportions according to a logic that is
part of our ancestral heritage, and has been proved by people who long ago

survived Pleistocene winters, the Ice Ages, great droughts, expeditions over land and sea, and persistent famines. Our genes remember our family history, and that once those who were fat were truly blessed.

21.2 Purpose

Obesity represents one of the most prevalent, diet-related medical problems in the United States, where approximately one-third of the population is over-weight. Given the high cost of treating obesity and the related disease states of hypertension, cardiovascular disease, and non-insulin-dependent diabetes mellitus, it seems prudent to develop tools for education that may contribute to prevention and treatment of the disorder. Obesity develops when a pro-longed, quantitative imbalance exists between energy intake and energy ex-penditure; prevention of obesity requires an understanding of this imbalance and modification of behaviors that predispose to weight gain. Because the problem is quantitative, it is well suited to computer simulation, and the com-puter should be used to help impart knowledge that can be useful in com-bating an intractable problem.

This chapter demonstrates one method to construct a simple model that il-lustrates the concept of fat and energy balance. The model permits the user to estimate initial body composition, resting and 24-h energy expenditure, energy balance, and 24-h fat oxidation rates based on published regression equations. Once a balance is established, the consequences of altering energy intake, physical activity, or fat intake can be tested in any combination.

This chapter also demonstrates that a plausible model may still produce er-roneous outcomes, and corrects a source of error in the original model. The completed model demonstrates that a fat deficit for a subject in energy bal-ance normally implies a carbohydrate excess, and the model overestimates weight loss unless energy from carbohydrates is allowed to make up for the deficit in fat.

21.3 Health-Related Goals Concerning Obesity

In developing a simulation model, it is important to define the purpose of the program before determining what features should be included. In the case of public-health-related goals concerning overweight or obesity, the objectives may be stated in terms of potential for contributing to goals published in *Healthy People 2000* (U.S. Department of Health and Human Services, 1990). Relevant goals were: Goal (1.2) Reduce overweight to a prevalence of no more than 20% among people aged 20 and older and no more than 15% among adolescents aged 12 through 19. Goal (1.3) Increase to at least 30% the proportion of people aged 6 and older who engage regularly, preferably daily, in light to moderate physical activity for at least 30 minutes per day. Goal (2.5) Reduce dietary fat intake to an average of 30% or less and average saturated fat intake to less than 10% of calories among people aged 2 and older. Goal (2.7) Increase to at least 50% the proportion of overweight people aged 12 and older who have adopted sound dietary practices combined with regular physical activity to attain an appropriate body weight. When considered together, these goals plainly require changes in behavior that include achieving a balance between food consumption and energy expenditure, reduced fat consumption, and increased physical activity. Therefore, the aim of this chapter is to create an interactive model that permits users (1) to establish an initial balance between food intake and energy expenditure; (2) to estimate current needs for food energy and grams of fat; (3) to predict the effect of reducing fat intake independently from reducing total energy intake; (4) to evaluate the effect of increasing physical activity; and (5) to predict the time course of weight gain or loss when an imbalance exists.

21.4 Assumptions and Sources of Equations

The model simulates changes in metabolizable fat mass (FM), which is considered to be triglyceride with an energy content of 38 kJ/g (9 kcal/g). The difference in weight between this functional definition of fat mass and the chemical definition of extractable fat was assumed to be negligible. Fat-free mass (FFM) was assumed to include all tissue constituents other than FM, and was assumed to contain 6 kJ/g (1.4 kcal/g). The model is intended to simulate long-term changes in weight as a result of altered fat or energy balance and assumes that changes in average carbohydrate stores are included in the change in FFM. The subjects being simulated are assumed to be nongrowing adults who are interested in weight maintenance or weight loss, because the energy cost of growth is higher than the energy cost of tissue maintenance. It is assumed that the energetic efficiency of deposition and lipolysis of triglyceride is 90% (this value may be changed in the model). The model assumes that an energy deficit produced a depression in metabolic rate that increased in proportion to the magnitude of the energy deficit to a maximum of 10% of the basal metabolic rate.

Regression equations for key relationships predicting metabolic rate and energy usage were obtained from the following sources: body composition as a function of weight, and resting metabolic rate predicted from FFM and FM (Nelson et al., 1992; Weinseir et al., 1993); change in lean mass as a function of a change in fat mass (Forbes, 1987); fat oxidation rate as a function of energy balance (Toubro et al., 1995).

The program functions in the following sequence: It estimates body fatness on the basis of the weight entered, and uses the quantity of fat mass and fat free mass to estimate resting energy expenditure (REE). A multiplier for physical activity is then applied to the REE, and an additional value for diet-induced thermogenesis (DIT) is added, equal to 10% of the energy intake. The quantity of fat metabolized by a person of these characteristics is then estimated on the basis of a regression equation derived from human en-

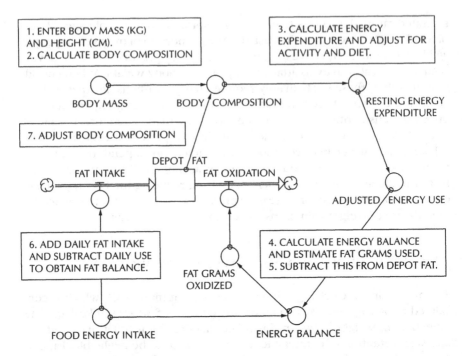

FIGURE 21.1

ergy balance studies. The fraction of total caloric expenditure estimated to be fat is converted to the predicted number of grams of fat oxidized each day. **STELLA®** software then subtracts the amount removed from the fat stores and adds the amount consumed per day on the basis of the dietary composition. If this results in a gain or loss of fat mass, the proportion of fat-free mass that would be gained or lost is calculated, and an adjustment is made. Thus, the program predicts changes in body composition as a function of food intake and activity level.

The general flow of logic in the model is shown in Figure 21.1.

21.5 How to Use the Program

One begins by using the sliders in the graphical user interface shown in Figure 21.2 to enter body mass (kg), height (m), and an activity level (1.4 is suggested). Fat mass is estimated by a regression equation, because most students will not have access to means of estimating fat mass; if it is known, it may be entered by opening the Fat Mass converter on the mapping level. To facilitate the initial data entry, a separate program allows one to convert values in pounds, inches, and kilocalories into Standard International units (2.2 pounds per kg, 2.54 cm/inch, and 4.18 joules/kcal). Then run the program briefly to estimate current energy expenditure. It is a good idea to estimate energy needs by this procedure, then enter the predicted value, and run

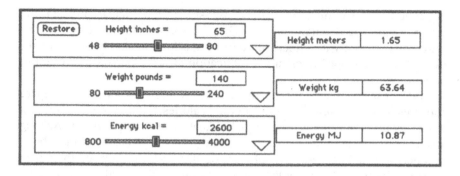

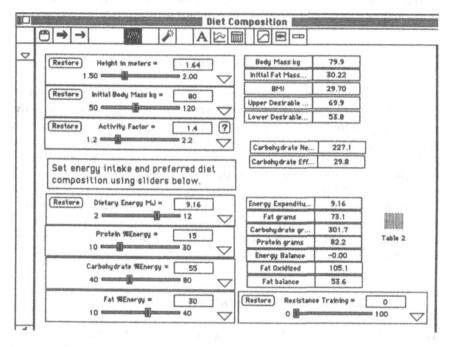

FIGURE 21.2

the program again. The time scale is set to 365 days, so that a full year is simulated; one may verify that an equilibrium has been established by showing that body weight does not change on this graph.

21.6 Tests to Perform Using the Fat Balance Program

Try simulating weight loss in an obese subject with a body mass index of about 30 or more. Once an energy and fat equilibrium is established, try reducing energy intake without modifying diet composition, and observe how long it takes to achieve a stable body weight. Try a similar test by increasing or decreasing the level of physical activity, and then combine the two. Having done that, try modifying the diet composition. Of particular interest is the effect of reducing fat intake to 30%, 25%, and 20% of calories. How does the outcome compare with effects of decreasing calorie intake, in other words, how much "leverage" does diet modification afford a person interested in weight loss?

Now consider a person who is interested in weight maintenance. Compare the effects of increasing the intake of total calories, fat, or carbohydrate by a small amount, equivalent to perhaps 100 kcal (0.418 MJ) per day. How long would it take to achieve a new, stable body weight? What would the body composition be? It is possible to plot any of the variables shown on the model level as a function of time, or list them in the tables. If desired, the data in the tables may be saved and plotted using graphics software.

21.7 A Source of Error in the Original Program

The original version of the fat balance program allowed a person to decrease fat intake at a given amount of dietary energy without showing how the body would compensate by using more carbohydrate or protein. This caused the person being simulated to lose weight, but people who change their diet without reducing their caloric intake do not generally lose weight. Therefore, a device was added that requires one to enter total dietary energy, a level of fat intake, and a level of protein intake. It then calculates the amount of carbohydrate needed to make up the fat deficit. The program determines whether the increased intake of carbohydrate or protein has created an excess relative to need. If there is an excess, it reduces the amount of fat oxidized, which is offset by use of the other nutrients.

21.8 Notes on Model Development

The model was constructed with a single compartment representing depot fat, with changes in lean tissue following passively according to the regression equation obtained by Forbes (1987). As fat mass increases, the amount of lean tissue added per kg of body weight decreases, so that body composition changes (Fig. 21.3). During weight loss, the opposite is true, and obese people lose more weight as fat than do lean people, who lose more lean. Daily fat intake in grams was entered as one of the initial conditions. Efficiency of fat deposition was assumed to equal 90–97% for the simulations presented here. Net lipogenesis was assumed to equal zero. The rate of depletion of depot fat was calculated from daily fat oxidation rates, which were assumed to depend exclusively on net energy balance. The regression equation that was derived from calorimetry studies of obese humans by Toubrou and colleagues (1995) was used to relate 24-h fat oxidation rates to 24-h energy balance [Fat fraction = 43.6% − 0.0032%/kJ * Energy balance (kJ)]. This produces a rate of fat oxidation equal to 50% of the nonprotein substrate

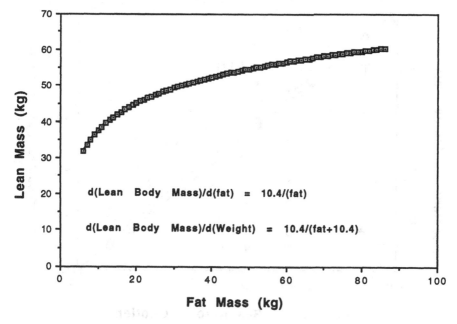

FIGURE 21.3. As human subjects gain weight, body composition changes and acquires a higher proportion of fat (Forbes, 1987). This graph indicates the relationship that was programmed into the **STELLA®** model described here.

oxidation for an energy deficit of −2.5 mJ (−600 kcal), about 44% in energy balance, and about 35% for an energy surplus of 2.5 mJ. The value at energy balance would correspond to a 24-h Respiratory Quotient (RQ) of 0.87, which is conservative. Figure 21.4 shows the relationship between the RQ and the amount of fat oxidized. A typical value for the 24-h RQ is 0.83, which indicates that about 60% of our caloric use (protein excluded) comes from fat. Meals increase insulin secretion, increase the use of carbohydrate, and decrease the use of fat; fasting has the opposite effect. Protein usually accounts for about 15% of the 24-h energy supply (Toubro et al., 1995), and the total amount of fat oxidized per day was estimated by multiplying the fat fraction of the nonprotein RQ by 85% of energy expenditure (EE).

21.9 Diet Composition and Conditions for Energy Balance and Fat Balance

Energy balance occurs when total dietary energy equals 24-h EE. At energy balance, the regression equation for fat use provides neutral fat balance when fat intake equals about 37% of 24-h EE, carbohydrate is 48% of EE, and protein is 15% of EE. At any specified energy intake, a reduction in dietary fat implies an increase in carbohydrate or protein intake. Therefore, reducing fat intake produces a positive balance in carbohydrate or protein. To model

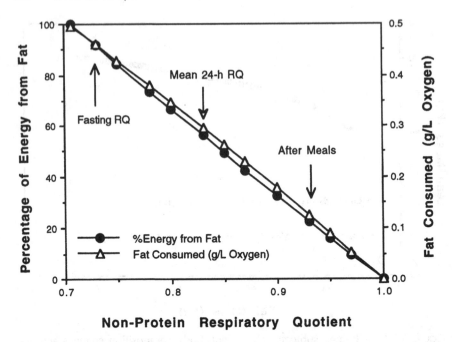

FIGURE 21.4. The respiratory quotient (RQ) can be used to estimate the amount of fat and carbohydrate being used by a subject. This relationship was used in the **STELLA®** model to estimate the amount of fat that would be oxidized by a subject as a function of energy balance (the difference between food energy intake and energy requirements).

this effect, the carbohydrate or protein balance was calculated: Each gram of positive carbohydrate balance reduced fat oxidation with an efficiency of 93%, and each gram of positive protein balance reduced fat oxidation with an efficiency of 76% (Devlin and Horton, 1990). These lower efficiencies of use for carbohydrate and protein are equivalent to subtracting a value for diet-induced thermogenesis when carbohydrate or protein are in excess of needs. Similarly, a negative balance in carbohydrate or protein relative to needs increased the amount of fat oxidation, and offset any increased intake of fat. Finally, alcohol was assumed to reduce fat oxidation with an efficiency of 74% (Murgatroyd et al., 1996).

21.10 Effects of Activity and Resistance Training

In order to calculate daily fat oxidation rates, an equation for resting energy expenditure (REE) that requires values for FM and FFM was used (Nelson et al., 1992). Initial conditions are set assuming that a subject's body composition is known; if not, body composition is estimated based on an equation provided by Weinseir and colleagues (1993): FM = 69.88 + 1.11 W −67.5 log W, where W equals body weight in kg. To calculate 24-h energy expenditure (EE), REE was multiplied by a value for the activity factor (AF). The value for AF was not assumed to be constant; the simulation program allows AF to be set to any value between 1.2 and 2.2. Finally, a value for diet-induced thermogenesis (DIT) equal to 10% of dietary energy was added to the adjusted REE. Energy balance was then calculated by comparing 24-h food intake with 24-h EE. This provided an estimate of the 24-h fat oxidation rate in units of mJ, which was converted to an equivalent number of grams of fat assuming an energy content of 38 kJ/g (9 kcal/g). During a simulation, the value for 24-h fat oxidation is calculated and then subtracted from the existing quantity of depot fat; a corresponding amount of FFM is then subtracted, and the next iteration begins.

The anabolic effect of resistance training on muscle was simulated by assuming that muscle mass was 60% of FFM, and could increase up to 10% if exercised at the equivalent of 3 workouts per week, with three sets of 12 lifts to produce a full effect. The half-time of the effect was assumed to be 2 weeks. The anabolic effect changes the ratio of fat:lean tissue lost during any specific diet, and increases the 24-h EE in proportion to the extra FFM. The energy cost of the weight training activity would need to be added by increasing the physical activity index.

List of Finite Difference Equations for One Set of Initial Conditions

```
Fat_Mass_kg(t) = Fat_Mass_kg(t - dt) +
(Fat_Intake_&_Synthesis - Daily_Fat_Oxidation) * dt
INIT Fat_Mass_kg = Initial_Fat_Mass_kg
Fat_Intake_&_Synthesis = (Fat_Intake_g_per_day +
Lipogenesis) * 1/1000
Daily_Fat_Oxidation = Fat_Oxidized * 0.001
Activity_Factor = 1.4
Adjustment = Metabolic_Depression *
Basal_Energy_Expenditure
Basal_Energy_Expenditure = (0.0703 * Fat_Free_Mass_kg +
0.0197 * Fat_Mass_kg + 1.8069)
BMI = Body_Mass_kg/(Height_in_meters^2)
Body_Mass_kg = Fat_Mass_kg + Fat_Free_Mass_kg-
Glycogen_Depletion
Change_in_Energy_MJ = IF(TIME = 0) THEN (0)
ELSE(Change_in_Fat_MJ + Change_in_FFM_MJ)
Change_in_Fat_Mass = Fat_Mass_kg - Initial_Fat_Mass_kg
Change_in_Fat_MJ = Change_in_Fat_Mass * 38
Change_in_FFM = Fat_Free_Mass_kg - Initial_FFM + 0.2
Change_in_FFM_MJ = Change_in_FFM * 6
```

```
Change_in_Weight = Change_in_Fat_Mass + Change_in_FFM -
Glycogen_Depletion
DIT = Food_Energy_Intake_mJ * 0.1
Energy_Balance = Food_Energy_Intake_mJ -
Energy_Expenditure_mJ
Energy_Expenditure_mJ = DIT + (0.0703 *
Fat_Free_Mass_kg + 0.0197 * Fat_Mass_kg + 1.8069) *
Activity_Factor
Fat_Fraction = (43.6 + (1000 * Energy_Balance * -
0.0032))/100
Fat_Free_Mass_kg = 14.2 + 23.9 * LOG10(Fat_Mass_kg)
Fat_Intake_g_per_day = 132.9
Fat_Oxidized = ((Energy_Expenditure_mJ - Adjustment) *
Fat_Fraction)/(0.038)
Food_Energy_Intake_mJ = 8.6
Glycogen_Depletion = (Fat_Free_Mass_kg *
Metabolic_Depression * 0.4) * (1 - EXP(-0.35 * TIME))
Height_in_meters = 1.67
Initial_Body_Mass_kg = 80
Initial_Fat_Mass_kg = 69.88 + (1.11 *
Initial_Body_Mass_kg) - 67.5 *
LOG10(Initial_Body_Mass_kg)
Initial_FFM = Initial_Body_Mass_kg -
Initial_Fat_Mass_kg
Lipogenesis = If(Energy_Balance> = 0) Then (4) Else (0)
Lower_Desirable_Weight_kg = Height_in_meters^2 * 20
MJ_per_kg = IF(TIME = 0) THEN(0)
ELSE(Change_in_Energy_MJ/Change_in_Weight)
Percentage_Oxidized = (Fat_Oxidized * 0.1)/Fat_Mass_kg
Upper_Desirable_Weight_kg = Height_in_meters^2 * 26
Metabolic_Depression = GRAPH(Energy_Balance)
(-6.00, 0.0995), (-5.40, 0.09), (-4.80, 0.0805),
(-4.20, 0.069), (-3.60, 0.06), (-3.00, 0.05), (-2.40,
0.04), (-1.80, 0.03), (-1.20, 0.02), (-0.6, 0.01),
(-6.44e-16, 0.0005)
```

21.11 How Much Fat Do We Oxidize in a Day?

In a single-compartment model, the quantity of substance present at equilibrium equals the rate of intake plus production (units, quantity per unit of time) multiplied by the efficiency of deposition and divided by the fractional elimination rate (units, time^{-1}). A typical amount of fat ingested per day is about 100 grams, and an average human maintains fat stores of about 20 kg.

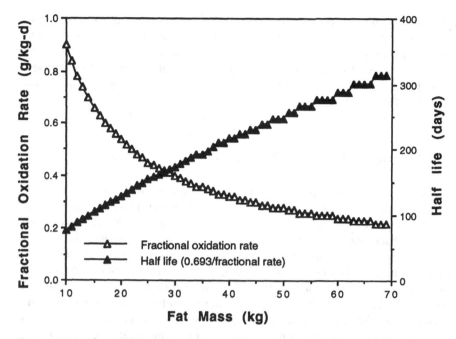

FIGURE 21.5. The relative rates of fat oxidation are compared for a range of body masses. As body mass increases, the daily rate of fat oxidation is expected to increase. However, as the individual becomes increasingly obese, the absolute fat oxidation rate represents an ever-smaller fraction of the total amount stored in depot fat. This fractional oxidation rate (open triangles) is a major determinant of the time required to mobilize fat stores. The projected half-life of stored fat is shown (closed triangles).

For a person in balance, then, the fractional elimination rate would equal $(0.1 \text{ kg d}^{-1}/20 \text{ kg}) = 0.005 \text{ d}^{-1}$. On an average day, 0.5% of fat stores would be eliminated by oxidation; this value represents *fractional usage* because it describes the fraction of the total depot fat present that will be eliminated each day.

It is interesting to divide the total amount of fat the body uses per day by the total amount of fat stores to estimate a fractional usage rate. Because lean tissue uses more energy than fat tissue and lean people have lower fat reserves, they use a larger proportion of their stored fat each day. The opposite is true for obese people (Fig. 21.5). In terms of weight maintenance, this confers an advantage on the lean person; in terms of having enough energy reserve to last through a time of hardship, it is disadvantageous to be overly lean.

The importance of this concept relative to weight loss is that the time needed to gain or lose fat depends on this fractional rate of use. A high usage rate allows one to modify body composition in less time than a low rate (Fig. 21.5). Note that the half-life, or time required to mobilize fat stores, progressively increases the fatter one becomes, and can be as long as a year in very obese people. It requires about 5 half-lives to achieve a new equilibrium, so it is evident that combatting weight gain can be very discouraging to people who find it easy to gain and hard to lose. It is physically impossible to achieve a permanent weight reduction of this magnitude in the few weeks usually promised by the nostrums sold in newspaper supplements. Students who understand that the First Law of Thermodynamics applies should wisely save their money for a program that does work, albeit slowly. The only truly rapid method is liposuction!

21.12 Effect of Energy Balance on Fat Oxidation

An energy deficit is expected to increase fat oxidation, as reflected in the 24-h respiratory exchange ratio, and the model reflects this behavior. This increase is offset to some extent because lowered food intake decreases diet-induced thermogenesis and causes a metabolic depression, and this causes total energy expenditure to decrease. Nevertheless, the change in fat oxidation as a result of fasting is relatively large. In this case, based on a 60 kg subject, the fat oxidation rate increased from 69 g/d to 112 g/d over a range of +3 to −6 MJ/d. This effect would be somewhat offset by a depression of the basal metabolic rate, but metabolic depression was not included in this model because the goal was to simulate effects of moderate changes in diet, not the effect of very low calorie diets.

21.13 Efficiency of Fat Deposition

The efficiency of fat deposition ranges from as low as 75% (50 kJ/g or 12 kcal/g) during tissue growth to as high as 97% during ordinary metabolism. This is a very important consideration, because efficiency is multiplied by fat intake (plus lipogenesis) to calculate the steady-state quantity of depot fat. Using the value that is typical of growth would predict that 75% of dietary fat was stored, which is 22% lower than the result of assuming a 97% deposition rate. The present simulations used a value of 97%; however, the efficiency may be changed simply by opening the dialog box associated with fat intake and changing the value to match the user's assumptions. It is important to note that the absolute amount of fat ingested is not expected to affect the time course to attain a new steady state; the time course is governed completely by the fractional oxidation rate, which effectively serves as a time constant for change.

21.14 Effect of Physical Activity on Fat Oxidation

The model captures the effect of physical activity as an increase in fat oxidation as a result of greater energy expenditure and, if food intake is not increased, as a result of a greater energy deficit. Activity, food intake, and fat intake are interdependent variables; each may be left at its initial value, increased or decreased, depending on the intentions of the modeler. As activity increases, total energy expenditure increases and fat oxidation increases in parallel.

21.15 Effect of Body Composition on Rate of Weight Change

For a subject in energy balance, the amount of fat oxidized each day depends on the 24-h energy expenditure. Therefore, the fractional oxidation rate depends on body composition and decreases as the proportion of FM increases relative to FFM. When the subject is not in steady state, the time required for the body composition to attain a new equilibrium is approximately 5 half-lives. Because obese individuals oxidize a smaller fraction of their fat depots each day than do lean subjects, the time required to attain a new equilibrium is greater for subjects who are obese (Fig. 21.5). The rate of approach to steady state is limited by the maximum daily rate of fat oxidation and the absolute size of the fat depot.

Because body composition changes as weight increases or decreases, the model predicts an asymmetry between the rate of weight gain and the rate of weight loss. With weight gain, the metabolic rate and the absolute amount of fat oxidized per day increase slower than body mass increases, whereas the opposite is true during weight loss. Therefore, the weight gain produced by a specified increase in energy intake is not equal to the weight loss produced by a decrease of the same magnitude. As pointed out by Weinseir and colleagues, the rate at which weight changes is also expected to differ for lean and obese people; weight gain and weight loss are both protracted in obese people because their existing fat depots are large and fractional turnover is small.

One advantage of the single-compartment model of FM is that it permits the steady-state fat mass to be estimated for any combination of fat intake, energy intake, activity, and body composition. In theory, the steady-state value equals the input rate (daily fat intake plus lipogenesis) divided by the fat oxidation rate. The fat oxidation rate for this calculation equals the number of grams of fat oxidized per day divided by the total number of grams of depot fat in the body. For example, consider a person with 30 kg of depot fat who ingested 100 g of fat per day and who required 7.5 mJ (1800 kcal) per day. If she derives 40% of her energy from fat, she metabolizes about $(0.4 \times 7.5)/0.0376 = 80$ grams of fat per day. Relative to her total depot fat, her daily rate equals $80/30{,}000 = 0.0027$. Therefore, her depot fat should increase until it equals $100/0.0027 = 37$ kg, and her weight should increase according to the relationship predicted by Forbes (1987).

The computer program predicts these changes automatically, and also estimates the time required to attain a new steady state. As pointed out by Weinseir and colleagues (1993), for a single-compartment model, the half-

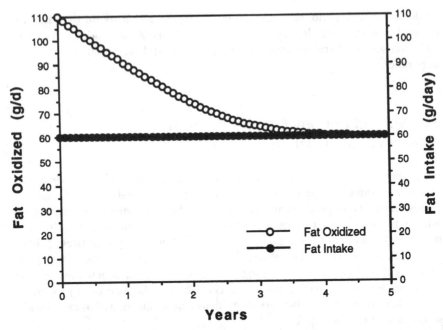

FIGURE 21.6. The model shows that body weight should be stable in energy balance. However, any change in energy intake will produce a change in fat oxidation (open circles) and, therefore, in body weight. The component in the system that adjusts slowest is the fat mass, which has a half-time of several months, and requires a period of years to adjust to a new equilibrium in which energy expenditure due to increased body mass now equals the increased energy intake.

time of approach to steady state in a single-compartment model is related to the fat oxidation rate (k). The half-time for change equals Ln 0.5/k, which equals 0.693/k. For the example given, 0.693/0.0027 = 256 days. A steady state is achieved after approximately 6 half-lives, so the equilibrium would not be achieved for about 4 years. However, the fractional rate is not a constant, as is required of a true single-compartment model; as fat mass increases, the fractional rate decreases. It is evident that there is no such thing as a steady state for human energy metabolism, and that the rates of change are depressingly sluggish.

Because of the nonlinear relationship between fat gain or loss and change in body weight, the model does not generate a constant energy cost per kg of weight lost (Fig. 21.6). Instead, an energy content of 37.6 mJ/kg FM and 7.24 mJ/kg FFM results in a range of about 24–30 mJ/kg weight (or 2,600–3,200 kcal/pound). These values are less than the rule of thumb that a pound of fat tissue contains about 32.2 mJ (3500 kcal) but similar to the value of 24.8 mJ (2700 kcal), which assumes that weight loss occurs at a ratio of 0.75 FM to 0.25 FFM.

21.16 Evidence That the Simplest Model of Fat Balance Was Wrong

Because the body oxidizes a significant amount of fat when in energy balance no matter what the composition of the diet, the postulate of mass balance leads to the idea that a diet low in fat ought to lead to loss of fat mass. Consider a test in which a subject who initially weighed 80 kg began to consume a diet containing only 30 g of fat but containing her predicted energy requirement of 9.6 MJ. The original model predicted that she would initially oxidize 112 g of fat per day, giving a fat deficit of 82 g/d. During a one-year period, her daily fat oxidation rate should decrease to a value of 75 g/d, yielding a cumulative loss of 22 kg of fat. Furthermore, this large weight loss would reduce her energy expenditure, so that by the end of the year, the subject would be in an energy surplus of 2 MJ/d while still losing weight! Needless to say, this prediction is incorrect!

21.17 Modifying the Model

A rule that all models of fat balance must follow is that only small amounts of weight will ever be lost when a person is taking in adequate energy. The reason for this is that the body's ability to derive energy by substituting one nutrient for another is great. When carbohydrate is limited, protein and amino acids will be used to provide glucose through gluconeogenesis, and fatty acids will be liberated from triacylglycerol. Similarly, excess carbohydrate releases insulin, which increases fat storage and augments use of carbohydrate for energy. This is evident after a meal, when the respiratory quotient rises above 0.9.

If this compensation did not occur, it would be possible to lose fat simply by reducing consumption of dietary fat. Although this is useful, it produces such a small effect compared to increased physical activity or decreased food intake that it hardly justifies the modern obsession with low-fat foods. To demonstrate this, the model was changed to allow any increased carbohydrate consumption to offset decreased fat consumption. The compensation was achieved by calculating needs for carbohydrate, fat, and protein based on intake of a balanced diet, with the sum of energy from all three nutrients equal to a fixed total of dietary energy. As fat consumption is reduced at any level of energy intake, one must specify how carbohydrate and protein are increasing.

The increased intake of carbohydrate or protein then offsets the need for fat. The efficiency by which carbohydrate calories may offset fat is approximately 90–95%; the value for protein is lower. These changes produced the new model that accompanies this text.

21.18 Different Ways to Use the Fat Balance Model

The model is intended to allow a person to compare the effects of changing food intake, diet composition, and aerobic activity on body weight and body composition, in order to evaluate different strategies for weight loss and weight maintenance. Under a variety of conditions, the model will estimate the amount of fat the body oxidizes per day, so that one does not need to allow 30% of calories or 60–65 grams of fat per day. Relative to this norm, it will then show how increasing consumption of carbohydrates or protein may reduce the use of fat.

The model can also be used as a guide to more general questions. For instance, it estimates the amount of fat stored for people of different weights and heights, and calculates the body mass index. At each weight, it calculates the amount of fat oxidized, so that one can observe the curve showing how lean people oxidize a large amount of fat relative to their stores, but the opposite is true for obese people. The model predicts a time course for weight loss or gain under various conditions, and demonstrates that the process is relatively slow for most people. It can truly require 1–2 years for an obese person to attain desirable weight despite a high level of commitment and good compliance with advice. The model can also show the effects of lack of compliance.

21.19 Major Conclusions

The major outcome of simulations is not surprising: the most important determinant of weight loss is the magnitude of the energy deficit, whether caused by decreased food intake or increased physical activity. Changes in weight due to reduction of fat intake are much more modest because individuals who select a low-fat diet usually increase their intake of carbohydrate or protein. Predictions of the model agree with the thesis that successful weight loss requires some reduction in energy from fat and carbohydrate, although modest effects of diet composition are observed. During energy restriction, the model predicts that weight will be lost as the amount of fat used per day gradually approaches the amount taken in from the diet; for anyone who is significantly obese, the process will take more than a year to complete (Fig. 21.6).

Although it is reasonable to model changes in depot fat and body composition using either a single- or dual-compartment model, it is evident that changes in fat mass do not follow such a simple model. For instance, the single-compartment model suggests that the reduction in depot fat caused by changes in diet or activity could be predicted by dividing the current fat intake by the fractional oxidation rate (which is equal to the amount of fat oxidized per day divided by the total fat mass). This idea is an oversimplification because any reduction in fat intake at a specified energy intake implies a corresponding increase in carbohydrate or protein, and the other nutrients are used in place of fat. This is evident from the rise in RQ that occurs if excess carbohydrate is ingested, and the decrease in RQ if fat is ingested. Because the other nutrients reduce fat oxidation, the effective fractional oxidation rate decreases. Thus, the new steady-state value would need to account for the change in elimination rate, and for the additional complication that as body weight decreases, the amount of fat required to support daily metabolism decreases. Although there may be no analytical solution to this problem, the model presented here allows the effects of changes to be estimated quickly.

The great advantage of modeling is that it allows scenarios to be tested and outcomes of choices to be observed in the manner of thought experiments. This allows key concepts to be reinforced, and captures the flavor of scientific experimentation. The model presented here was designed to capture a number of key elements of human physiology. For example, it allows one to examine what happens when energy intake and energy expenditure are balanced, when food intake exceeds needs, or when intake is inadequate. The program estimates amount of fat oxidized each day, so that the student can compare intake and usage under a variety of conditions; this is a concept that is very poorly developed in most people. By providing an estimate of the size of the fat depot relative to daily needs and dietary intake, the program can help the student understand why significant weight loss takes a long time, and why the daily increment in depot fat may be so small that few people even notice that a problem may be developing. Even with a small change in food intake, several years are required to achieve a new weight at which energy usage now equals energy intake. More advanced students might use the program to think about the relative stability of body composition, the concept of the steady state, and the physical forces that govern the time course of change. By showing effects of diet and activity on body composition, the program can aid in the concept that excessive loss of lean tissue is undesirable and that a sound plan for weight maintenance preserves lean tissue.

One of the major predictions of the complete model is that changes in dietary fat intake alone have little effect on weight loss if energy intake is kept constant. Instead, the major determinant of a change in weight is the magnitude of the energy deficit or surplus, whether achieved by dietary restriction or altered physical activity. As the model was being developed, it was observed that a model based strictly on the fat balance concept produced unreasonable results if fat oxidation was treated as being separate from carbohydrate supply. The results differed from the commonplace observation that subjects who are in energy balance should neither gain nor lose weight, irrespective of the composition of the diet. Despite this, there is good evidence that human subjects who consume a high-carbohydrate, low-fat diet continue to oxidize fat. For example, Hill and colleagues (1991) observed that subjects fed a diet with 60% of calories from carbohydrate and 20% from fat (FQ = 0.917) for one week had an average RQ of 0.856, which indicates that 46% of the nonprotein calories came from fat. The subjects were in a negative fat balance equivalent to about 1.6 MJ/d, which would equal about 40 g of fat. This is equivalent to a daily fat oxidation rate of about 80 g, of which half came from the diet. These values are similar to predictions from the model, and would yield a loss of 14.6 kg of fat per year if sustained.

References

Acheson, K.J., Y. Schutz, T. Bessard, K. Anantharaman, J.P. Flatt, and E. Jequier. "Glycogen storage capacity and de novo lipogenesis during massive carbohydrate overfeeding in man." *Am. J. Clin. Nutr.* 48 (1988): 240–247.

Devlin, J.T., and E.S. Horton. "Energy requirements.", In: M.L. Brown, ed., *Present Knowledge in Nutrition*, 6th ed., Washington, D.C., International Life Sciences Institute, 1990, 1-6.

Flatt, J.P. "The differences in the storage capacities for carbohydrate and for fat, and its implications in the regulation of body weight." *Ann. N.Y. Acad. Sci.* 499 (1987): 104–123.

Flatt, J.P. "Dietary fat, carbohydrate balance, and weight maintenance." *Ann. N.Y. Acad. Sci.* 683 (1993): 122–140.

Flatt. J.P. "Use and storage of carbohydrate and fat." *Am. J. Clin. Nutr.* 61 (suppl. 1995): 952S-959S.

Forbes, G.B. "Lean body mass-body fat interrelationships in humans." *Nutr. Rev.* 8 (1987): 225–231.

Girardier, L. "L'autoregulation du poids et de la composition corporelle chez l'homme. Une approche systemique par modelisation et simulation." *Arch. Int. Phys. Biochim. Biophys.* 102 (1994): A23–A35.

Hargrove, J.L. *Kinetic Modeling of Gene Expression*, Austin, Tex.: RG Landes Company, 1994.

Hill, J.O., J.C. Peters, G.W. Reed, D.G. Schlundt, T. Sharp, and H.C. Greene. "Nutrient balance in humans: Effects of diet composition." *Am. J. Clin. Nutr.* 54 (1991): 10-17.

Maron, M.J. and R.J. Lopez. *Numerical Analysis: A Practical Approach*. Belmont, Ca.: Wadsworth Publishing Co., (1991), 423-497.

Murgatroyd, P.R., M.L. VanDeVen, G.R. Goldberg, and A.M. Prentice. "Alcohol and the regulation of energy balance: overnight effects on diet-induced thermogenesis and fuel storage." *Br. J. Nutr.* 75 (1996): 33–45.

National Research Council. *Recommended Dietary Allowances*, 10th edition. Washington, D.C.: National Academy Press, 1989.

Nelson, K.M., R.L. Weinseir, C.L. Long, and Y. Schutz. "Prediction of resting energy expenditure from fat-free mass and fat mass." *Am. J. Clin. Nutr.* 56 (1992): 848-856.

Neubig, R.R. "The time course of drug action." In: Pratt, W.B., and P. Taylor (eds.) *Principles of Drug Action. The Basis of Pharmacology*, 3rd ed. New York: Churchill Livingstone, 1990, 297-363.

Oritsland, N.A. "Starvation and survival and body composition in mammals with particular reference to Homo Sapiens." *Bull. Math. Biol.* 52 (1990): 643-655.

Payne, P.R., and A.E. Dugdale. "A model for the prediction of energy balance and body weight." *Ann. Hum. Biol.* 6 (1977): 525-535.

Peterson, S, and B. Richmond. *STELLA® II Technical Documentation*, Hanover, N.H.: High Performance Systems, 1993.

Schutz, Y. "The adjustment of energy expenditure and oxidation to energy intake: the role of carbohydrate and fat balance." *Int. J. Obesity* 17 (Suppl. 3, 1993): S23–S27.

Schutz, Y., A. Tremblay, R.L. Weinseir, and K.M. Nelson. "Role of fat oxidation in the long term stabilization of body weight in obese women." *Am. J. Clin. Nutr.* 55 (1992): 670–674.

Swinburn, B., E. Ravussin. "Energy balance or fat balance?" *Am. J. Clin. Nutr.* 57 (Suppl., 1993): 766S–771S.

Toubro, S., N.J. Christensen, and A. Astrup. "Reproducibility of 24-h energy expenditure, substrate utilization, and spontaneous physical activity in obesity measured in a respiration chamber." *Int. J. Obesity* 19 (1995): 544–549.

U.S. Dept. of Health and Human Services, Public Health Service. *Healthy People 2000: National Health Promotion and Disease Prevention Objectives.* Washington, D.C.: U.S. Government Printing Office, 1990.

Weinseir, R.L., D. Bracco, and Y. Schutz. "Predicted effects of small decreases in energy expenditure on weight gain in adult women." *Int. J. Obesity* 17 (1993): 693–699.

Westerterp, K.R., J.H.H.L.M. Donkers, E.W.H.M. Fredrix, and P. Boekhoudt. "Energy intake, physical activity and body weight: a simulation model." *Br. J. Nutr.* 73 (1995): 337–347.

22

Human Cholesterol Dynamics

Nothing could save me and my liver, if I were in charge. For I am, to face
the facts squarely, considerably less intelligent than my liver.

Lewis Thomas,
The Lives of a Cell, 1974, p. 66.

In subjects as varied as pilot training and and principles of cardiopulmonary
resuscitation, simulation is used to provide information and experience to
students prior to their attempts to fly an airplane or resuscitate a person
whose heart has stopped beating. How effective is dynamic modeling in en-
hancing the analytical and predictive skills of students? The question may still
be open. Nonetheless, it is a form of active learning that provides a novel way
of thinking and learning (Dunkhase and Penick, 1991; Modell and Michael,
1993). Ideally, techniques used to promote active learning should engage students in building conceptual models that can be validated with real data and
used to test hypotheses (Richmond and Peterson, 1993). The resulting feedback improves retention of information and enhances the ability to use new
information in other contexts (Felder, 1993; Michael, 1993). The purpose of
this chapter is to adapt dynamic models of cholesterol metabolism in humans
to an educational setting.

Mathematical modeling has been used by nutrition scientists to analyze the
biodynamics of vitamins, minerals, cholesterol, and lipoproteins (Jacquez,
1985; Green and Green, 1990; Collins, 1992). Essentially, models quantify the
intake, use, and exchange of nutrients among interlinked, kinetic compartments. For example, a rapidly exchangeable pool of cholesterol can be observed that is probably derived from cholesterol in various lipoproteins as
well as the liver and other tissues (Goodman et al., 1980). Modeling incorporates several unique features that are valuable in education as well as in scientific studies. First, models are based on quantities of nutrients that are actually ingested and observed in human tissues during dietary assessment.
Second, modeling requires knowledge of measured rates of intake, use, and
excretion, just as in formulation of the recommended dietary allowances.
Therefore, simulation can reflect true rates of change during normal health,
in deficiencies, toxic conditions, or in disease. Third, negative feedback can
be incorporated into models, such as the effect of dietary cholesterol on cholesterol synthesis, or the down-regulation of the low density lipoprotein
(LDL) receptor by saturated fatty acids.

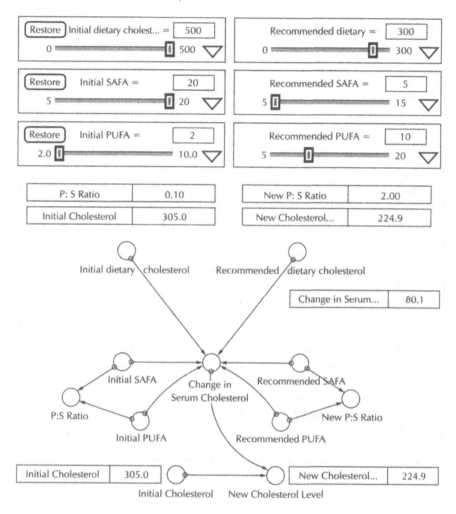

FIGURE 22.1

22.1 Effects of Dietary Cholesterol and Fat on Serum Cholesterol

STELLA® can be used to solve simple equations related to diet and cholesterol. Hegsted (1991) published an equation relating changes in dietary saturated and polyunsaturated fats and cholesterol intake to changes in serum cholesterol. The first model, shown in Figure 22.1, allows students to solve this equation to test expectations concerning effects of dietary change in responsive individuals.

Equations for the Hegsted Model

```
Serum_Cholesterol(t) = Serum_Cholesterol(t - dt) +
(Diet - Elimination) * dt
INIT Serum_Cholesterol = 340
INFLOWS:
Diet = 300
OUTFLOWS:
Elimination = Serum_Cholesterol * 0.01
Change_in_Serum_Cholesterol = 2.16 * (Initial_SAFA -
Recommended_SAFA) - 1.65 * (Initial_PUFA -
Recommended_PUFA) + 0.168 *
(Initial_dietary_cholesterol -
Recommended_dietary_cholesterol) + 0.857
Initial_Cholesterol = 305
Initial_dietary_cholesterol = 300
Initial_PUFA = 5
Initial_SAFA = 15
New_Cholesterol_Level = Initial_Cholesterol -
Change_in_Serum_Cholesterol
New_P:S_Ratio = Recommended_PUFA/Recommended_SAFA
P:S_Ratio = Initial_PUFA/Initial_SAFA
Recommended_dietary_cholesterol = 300
Recommended_PUFA = 10
Recommended_SAFA = 5
```

To use this model, set the initial levels of cholesterol and fat intake to unfavorable levels, such as a cholesterol intake of 400 mg per day or more, a high level of saturated fat, and a low level of unsaturated fat. Then set "recommended" levels of cholesterol to an intake not over 300 mg per day (it is possible to take in zero cholesterol), and increase the intake of polyunsaturated fat to a level of 7–10% of total daily calories, and decrease the saturated fat to a level not over 10% of total calories. Run the program and observe the values shown in the numeric displays.

The predicted outcome is not observed in many nonresponsive patients. Nor does this model suggest how long a period of treatment might be required to observe the predicted change. The only significant path for cholesterol elimination is via the bile, and the rate is not high due to efficient reabsorption and reuse of secreted cholesterol. Therefore, a kinetic model was also devised. The simplest one is described next.

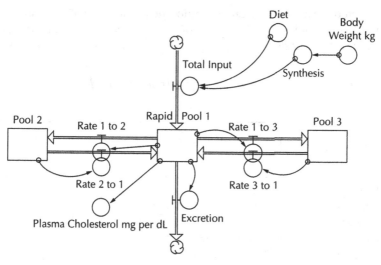

FIGURE 22.2

22.2 Minimal Model of Cholesterol Metabolism

A quantitative model of cholesterol dynamics was created using **STELLA®** modeling software based on studies of cholesterol kinetics from the clinical literature. The **STELLA®** diagram in Figure 22.2 shows the simplest model of human cholesterol metabolism that accurately reflects human cholesterol kinetics (Goodman et al., 1980). The diagram represents a rapidly exchangeable pool of cholesterol labeled 1 and two slowly exchangeable pools. No attempt is made to identify these with a particular organ in the body. The daily cholesterol production rate, which combines dietary intake and biosynthesis, is distributed among the three compartments, and an elimination rate is derived from the rapidly exchangeable pool. Cholesterol exchanges among the compartments at rates that are expressed as the fraction of material that exchanges in each direction per day. For instance, the elimination rate in normal subjects averaged 0.046 d^{-1}, which means that 4.6% of the cholesterol in the rapidly exchangeable pool is eliminated each day. This model was translated into a functioning computer program using **STELLA®** software and is included with this book. To simplify using the program, the dietary input and rate of synthesis were separated, and an equation relating body weight to rate of synthesis was included.

The rates and quantities of cholesterol used in this model are: pool 1, 24 g (of cholesterol); pool 2, 16.7 g; pool 3 32.7 g; absorbed dietary cholesterol, 0.2 g d^{-1}; synthesized cholesterol, 0.9 g d^{-1}; elimination rate, 0.046 d^{-1}; transfer rate from 1 to 2, 0.058 d^{-1}; rate from 2 to 1, 0.083 d^{-1}; rate from 1 to 3, 0.026 d^{-1}, rate from 3 to 1, 0.019 d^{-1}. A more detailed model that also incorporated the effect of dietary P:S ratio, the partitioning of cholesterol into LDL and HDL particles, and the effects of bile acid sequestrants and inhibitors of cholesterol synthesis, was developed as described (Foss et al., 1994).

Equations for the Three-Compartment Model

```
Pool_2(t) = Pool_2(t - dt) + (Rate_1_to_2 -
Rate_2_to_1) * dt
INIT Pool_2 = 9.73
INFLOWS:
Rate_1_to_2 = Rapid___Pool_1 * 0.058
OUTFLOWS:
Rate_2_to_1 = Pool_2 * 0.083
Pool_3(t) = Pool_3(t - dt) + (Rate_1_to_3 -
Rate_3_to_1) * dt
INIT Pool_3 = 19
INFLOWS:
Rate_1_to_3 = Rapid___Pool_1 * 0.026
OUTFLOWS:
Rate_3_to_1 = Pool_3 * 0.019
Rapid___Pool_1(t) = Rapid___Pool_1(t - dt) +
(Total_Input + Rate_3_to_1 + Rate_2_to_1 - Excretion -
Rate_1_to_3 - Rate_1_to_2) * dt
INIT Rapid___Pool_1 = 13.9
INFLOWS:
Total_Input = Diet + Synthesis
Rate_3_to_1 = Pool_3 * 0.019
Rate_2_to_1 = Pool_2 * 0.083
OUTFLOWS:
Excretion = Rapid___Pool_1 * 0.046
Rate_1_to_3 = Rapid___Pool_1 * 0.026
Rate_1_to_2 = Rapid___Pool_1 * 0.058
Body_Weight_kg = 74
Diet = 0
Plasma_Cholesterol_mg_per_dL = Rapid___Pool_1 * 8.3
Synthesis = 0.02 * (Body_Weight_kg - 28)
```

Model Capabilities

The basic model depicted in Figure 22.2 was used to explain the relative effects of changing dietary intake, the rate of elimination, and body weight. Because endogenous synthesis is typically near 1 g per day, reducing dietary cholesterol has an important but limited impact. The basic model did not include feedback suppression of cholesterol synthesis, and may overestimate the effect of changing intake alone. However, an increase in elimination rate obtained by increasing dietary fiber or use of bile acid sequestrants produced an additive effect, and therefore lowered cholesterol further. Because the rate of cholesterol synthesis increases with body weight and increases when dietary intake is lowered, it is difficult for an obese individual to reduce the hyperlipidemia by decreasing cholesterol intake alone. When weight reduction was combined with a low-cholesterol, high-fiber diet, the maximal result was achieved. The model clearly demonstrates the efficacy of combined dietary therapies.

Figure 22.3 shows effects of dietary therapies on plasma cholesterol as predicted using the kinetic model of human cholesterol metabolism. With no intervention, plasma cholesterol remained at 210 mg/dL (open circles). Eliminating all sources of dietary cholesterol reduced plasma cholesterol (closed circles). Combining a high-fiber diet with cholesterol reduction decreased plasma cholesterol further (open squares). The greatest reduction is predicted when these therapies are combined with a weight loss regimen (closed triangles).

It is important for students and counselors to understand how soon a given therapy is expected to produce an effect, and this is predictable from the kinetic model on the basis of the elimination rate. An elimination rate of 5% per day corresponds to a half-life of 0.693/0.05, or 14 days. Because it takes about 6 half-lives to achieve a new equilibrium, it is predicted that the maximum

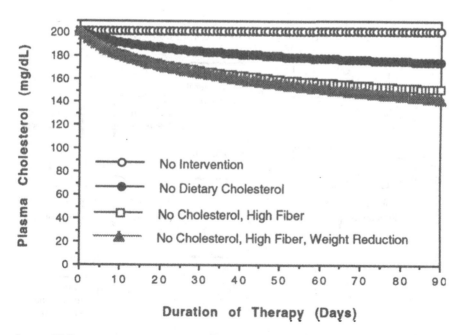

FIGURE 22.3

effect will occur after about 2–3 months of compliance with therapy. The simple model can be used to test potential effects of changing the diet to reduce cholesterol intake, but does not address more complex issues related to dietary saturated fat and fiber intake or the complexities of lipoprotein metabolism. For this reason, a more complex model was developed that provides a user interface to enable these questions to be addressed.

22.3 The Complex Model of Cholesterol Dynamics

Figure 22.4 shows an outline of cholesterol metabolism. Foss and colleagues (1994) generated a model of cholesterol metabolism based on the information shown in this diagram for the purpose of teaching dietetics students about effects of diet and medications on cholesterol metabolism.

Due to the complexity and size of the developed model, a High-Level map was added to house the simulation input and output controls for the complex model. In the control panel, the allowable inputs are weight, dietary cholesterol ingestion, dietary P:S (polyunsatured fatty acids/saturated fatty acids) ratio, and intake of cholesterol-lowering medications.

Rather than explaining the model in the text, it is suggested that the model be opened and used as follows. Open the High Level map and begin with a body weight of 70 kg, a cholesterol intake of 200–300, and a P:S ratio of 0.25. Leave the settings for the two medications (a bile acid sequestrant and an inhibitor of cholesterol synthesis) set at values of 1.0 (this gives expected lev-

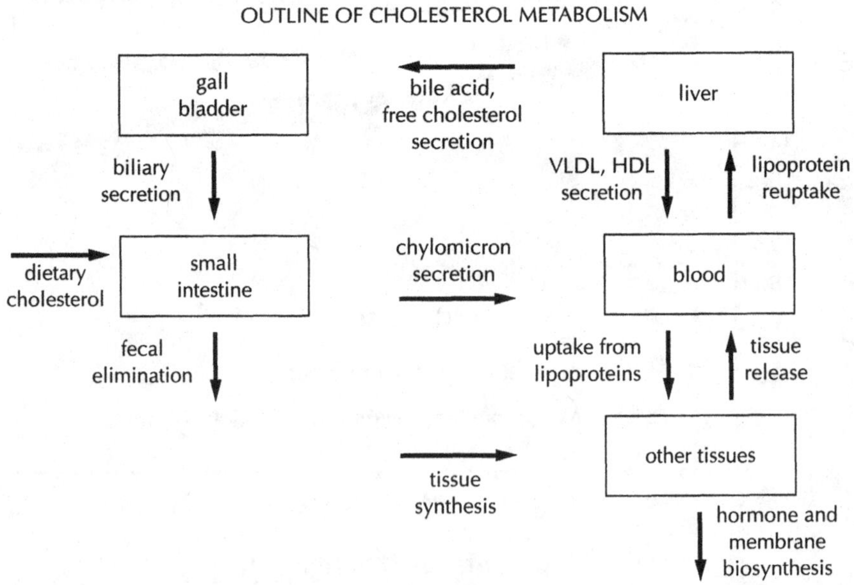

OUTLINE OF CHOLESTEROL METABOLISM

FIGURE 22.4

els of cholesterol; setting the switch to 2.0 indicates use of medication and will reduce cholesterol levels). Initial values in the program are set at 130 mg VLDL/LDL cholesterol, 40 mg/dL HDL cholesterol, and an exchangeable hepatic pool of 1610 mg. After performing a trial run, increase the cholesterol intake and modify the P:S ratio, and observe the effects. The numerical displays include HDL cholesterol and the sum of VLDL plus LDL cholesterol. If desired, add a display for total cholesterol. The beginning values for healthy people should be 140–200 mg/dL. Thus, the model is set to show the effects of adverse dietary changes or various preventative measures. If one were interested in beginning with a patient with high cholesterol, it would be necessary to allow the model to come into equilibrium at higher intakes and change the initial values in the different pools to reflect the new steady state before testing changes.

References

Collins, J.C. "Resources for getting started in modeling." *J. Nutr.* 122 (1992): 695–700.

Dunkhase, J.A., and J.E. Penick. "Problem solving for the real world." *J. Coll. Sci. Teach.* 21 (1991): 100–104.

Felder, R.M. "Reaching the second tier—learning and teaching styles in college science education." *Jour. Coll. Sci. Teach.* 22 (1993): 286–290.

Foss, J.D., D.K. Hartle, and J.L. Hargrove. "Dietary and pharmacological effects on blood cholesterol level: simulation of human inter-organ cholesterol metabolism." In: Anderson, J.G. and M. Katzper, eds. *Simulation in the Health Sciences.* Proceedings of the 1994 Western Multiconference. San Diego: Society for Computer Simulation, 1994, 60–65.

Goodman, D.S., F.R. Smith, A.H. Seplowitz, R. Ramakrishnan, and R.D. Dell. "Prediction of the parameters of whole body cholesterol metabolism in humans." *J. Lipid Res.* 21 (1980): 699–713.

Green, M.H., and J.B. Green. "The application of compartmental analysis to research in nutrition." *Ann. Rev. Nutr.* 10 (1990): 41–61.

Hegsted, D.M. "Dietary fatty acids, serum cholesterol, and coronary heart disease." In: Nelson, G.J., ed. *Health Effects of Dietary Fatty Acids*, Champaign, Ill.: American Oil Chemists Societes, (1991), 50–68.

Jacquez, J.A. *Compartmental Analysis in Biology and Medicine.* Ann Arbor: The University of Michigan Press, 1985.

Michael, J.A. "Teaching problem solving in small groups." *Ann. N.Y. Acad. Sci.* 701 (1993): 37–48.

Modell, H.I., and J.A. Michael. "Promoting active learning in the life science classroom: defining the issues." *Ann. N.Y. Acad. of Sci.* 701 (1993): 1–7.

Richmond, B., and S. Peterson. *STELLA®, An Introduction to Systems Thinking.* Hanover, N.H.: High Performance Systems, Inc., 1993.

Swetz, F. and J.S. Hartzler. *Mathematical Modeling in the Secondary School Curriculum.* U.S.: The National Council of Teachers of Mathematics, Inc., Reston, Va 1991.

23

Stochastic Model of Bone Remodeling and Osteoporosis

The race is not to the swift, but that's the way to bet.

Damon Runyon

Osteoporosis is a condition of skeletal fragility manifested by an increased risk of bone fracture that is now a major health problem in many countries (Melton, 1991). Oddly, this bone disorder is a sign of successfully living a long life, for it is relatively uncommon among people less than 50 years of age. Nevertheless, it is crippling to those afflicted, and there is a sense of paradox in that osteoporosis is more readily prevented than cured. Any person with risk factors for osteoporosis needs to become mindful that preventative strategies should begin in adolescence, or at the earliest possible stage in life, for the greater the bone density achieved during one's growing years, the longer it will be before bone deterioration begins to affect health.

Osteoporosis results in a humped posture (kyphosis), chronic pain, pressure sores resulting from needed bed rest, and decreased mobility (Wardlaw, 1993). Risk factors for this disorder include being of European or Asian descent, a family history of small or lean build, nulliparity, allergy to milk or long-term low-calcium diet, cigarette smoking, excessive alcohol consumption, a sedentary lifestyle, and long-term use of corticosteroids or thyroxine (Hillner, 1992). Due to the large number of risk factors and the fact that all women are susceptible to osteoporosis after the menopause, this disorder affects millions of individuals past the age of 50. Men are not completely protected, for any man who lives past the age of 70 has significantly decreased bone mass.

The original purpose of the model developed in this chapter was to provide an interactive means of teaching a unit on bone remodeling in a graduate-level nutrition course. In developing the model, something unique happened that I can only describe as a paradigm shift. My student, Ms. Kim Smith Porter, began the project with the aim of developing a model of calcium metabolism akin to one published by Jaros and colleagues (1980), which was a dynamic model of endocrine control of serum calcium and bone mass. After she completed the model and we were validating it, I observed the 1.2 kg mass of calcium in the subject's skeleton, and asked Kim, "So, what does this have to do with bone trabeculae?" She looked at me somewhat glumly, having just spent six months in an exhaustive literature review. "Go back to the literature," I implored, "and review any work based on bone remodeling, not the endocrinology of vitamin D and parathyroid hormone!" For the first concept nutrition students learn concerning calcium metabolism is the role of vitamin D in preventing rickets, simply because all nutrition texts introduce the fat-soluble vitamins before the minerals. However, the fundamental problem in osteoporosis is based on the remodeling cycle that occurs in bone trabeculae, which is only indirectly related to feedback regulation of serum calcium. Remodeling is a microscopic, stochastic event, whereas regulation of calcium homeostasis is a systemic process. I am pleased to report that the outcome of Ms. Porter's work has changed our curriculum, for osteoporosis is a vastly more important health problem than is rickets or the adult form of the disease, osteomalacia.

The stochastic model of bone remodeling included here deserves special comment. Ms. Porter included every pertinent reference from her literature review in the documentation that is embedded in the model. In a sense, the model represents the entire outcome and contents of her thesis, except for the separate model of calcium endocrinology. There is a stochastic model of trabecular loss; a high level map that includes a pictorial tutorial that explains the remodeling cycle; and an extensive list of sources, references, and reasons for choosing particular values. For this reason, it would be redundant to reproduce that material in this text. Instead, let us just explain one way of using the model. Anyone who wishes to learn the capabilities of **STELLA**® should examine this model; I hope you too will be impressed by what this fine student accomplished. She also taught other graduate students to use the model as part of a unit on bone metabolism, and the other students were very enthusiastic about this alternative approach to learning about bone.

Finally, every educator now has the capability to use computer-based projection systems in the classroom. I encourage any reader to open this model and use it while considering its applications either in a hands-on computer laboratory or as supporting material for a unit dealing with bone metabolism. The meaning of stochastic processes becomes very evident in this model, for a student will never observe the same outcome twice! This allows students to visualize the dynamics in a unique and dramatic manner, and understand that potential outcomes may vary for each individual. The next section was modified from the worksheet that was used to explain the model in conjunction with the graduate class.

23.1 Computer Simulation of Bone Remodeling and Osteoporosis as a Tool for Medical Education

What is osteoporosis? Osteoporosis may be broadly defined as a disorder in which "small modifications of bone remodeling give risk to substantial losses after a long period of time" (Delmas, 1992). Thus, it is the properties of the bone remodeling unit, the coupled resorption-formation process, that determine bone density maintenance. Normal trabecular bone consists of a meshwork of horizontal and vertical components of trabecular bone. It has been observed that osteoporotic trabecular bone lacks many of the horizontal components, which are vital to bone strength. At the present time, it is thought that the loss of bone trabeculae, such as the horizontal components, is irreversible. No treatment will restore bone to its normal, preosteoporotic configuration, thus health care professionals must pursue public preventive education in order to minimize the incidence of osteoporosis. (After reading this, it would be appropriate to open the model and navigate among the diagrams and photomicrographs located in the high level map).

Because bone loss is heterogenous throughout the skeleton and vertebral fractures are typically the first evidence of osteoporosis, we have chosen to simulate changes in the horizontal trabeculae in the third lumbar vertebra during the course of the aging process. We have assumed that a finite number of horizontal trabeculae exist within the third lumbar vertebra. The working model (located in the modeling level in the **STELLA®** routine) is divided into four sectors:

A. Control Panel

The control panel contains the majority of data needed to run the simulation. Basically, the control panel contains normally distributed data in reference to the third lumbar vertebra, including parameters of horizontal trabecular thickness, minimum critical thickness necessary for survival, and osteoclastic resorption depth.

B. Activation and Resorption

Some horizontal trabeculae are always being recruited (activation) to undergo the remodeling process. Trabeculae in the resorption process are randomly assigned numerical values of initial thickness, resorption depth, and minimum critical thickness from the control panel. In the process of resorption, trabeculae are permanently lost from the simulation by either (1) perforation by a resorption depth greater than the assigned initial thickness, or (2) failing to meet the criteria of critical thickness necessary for survival. Those trabeculae surviving the process of resorption go on to reversal.

C. Reversal

Reversal is a period of rest for the trabeculae being remodeled.

D. Formation

The trabeculae undergo the formation process one at a time. Each trabecula undergoing the formation process is assigned a unique formation time based on a normal distribution and then returns to the pool of quiescent trabeculae.

Now that you understand the basis of the simulation, let's begin exploring.

23.2 The Three Levels of the Stochastic Model

1. High Level Map

The high level map is an overview of the key players in the simulation. Notice the sliders, which are the red rectangular boxes, entitled Starting Age, Menopausal Act Freq, and Menopausal Res Depth. The activation frequency is defined as the number of remodeling events that begin in a specified period, and the resorption depth is the depth of the trabecular bone that is eroded by osteoclasts. We will be running our simulations from this layer. Move the cursor to the blue box, Activation and Resorption. Double click the mouse to bring up the text contained within Activation and Resorption. Use the up and down arrows on the keyboard to scroll through the text. Each icon in the high level map contains a text that may be perused at your leisure. To close the text, click the mouse once on OK. To move down to the next layer, click the mouse once on the down arrow in the left margin of the screen.

2. Diagram or Model Level

(Refer to Figure 23.1, or better yet, open the model.) The diagram layer consists of the constructed simulation and may be viewed in two modes. To change modes, click the mouse once on the globe in the left margin of the screen. The global mode contains a text about each icon, and the mathematical mode contains the equations necessary to run the simulation. As in the high level map, double clicking on any icon will bring up either a text or the equation for the icon, depending on what mode you have chosen. Notice the arrows pointing up and down in the left margin. Click the mouse once on the down arrow to proceed to the Equations layer.

3. Equations and Documentation

The third layer consists of the equations necessary to run the simulation. To go back up to the top layer, click the mouse once on the up arrow at the left margin. Click the mouse once on the top arrow again at the diagram layer to move back up to the high level map. (Note that in addition to the equations, this level includes the extensive list of publications from which Ms. Porter drew the information used in the model, and that she also explains her reasons for choosing most of the values.)

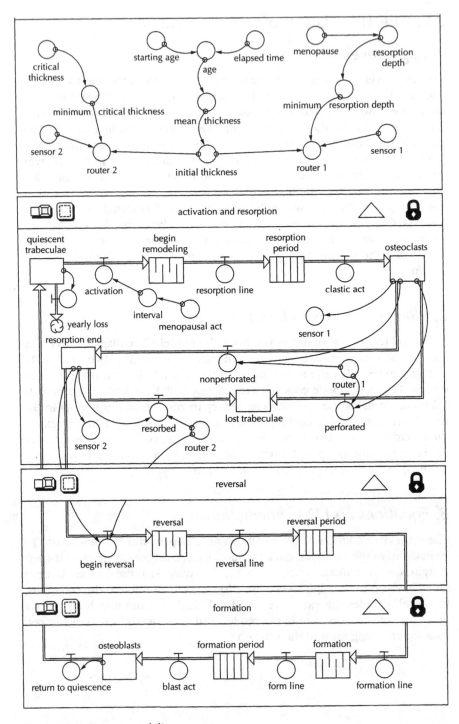

FIGURE 23.1. Bone remodeling.

23.3 Compare the Loss of Trabeculae in Young and Old Individuals

Return to the high level map and locate the control devices called sliders. These are used to set starting values for the model. One may either type in a specific number in the small box on the right side of each slider, or use the mouse to set the slider Starting Age to 20 years. Select the graph icon (light blue symbol on the right side of the menu) by clicking the mouse once. Deposit the graph on the high level map in any available space by clicking the mouse once. The control panel is depicted in Figure 23.2.

Next choose Define Graph from the menu under the heading Map and I/O. From the allowable variables, click the mouse once to select Initial Trabecular Thickness and Minimum Critical Thickness. Initial Trabecular Thickness represents the normal distribution of trabecular thickness, or width. Minimum Critical Thickness represents the normal distribution of trabecular thickness necessary for trabeculae to survive the remodeling process. Once you have done this, click the mouse once on OK. Then, choose Run from the menu and watch the results of your simulation. To preserve the graph, click the mouse once on the lock at the bottom left of the graph to "lock" your results into place. Close the graph by clicking the mouse once on the small rectangle at the top left of the graph. Follow the same procedure again for a

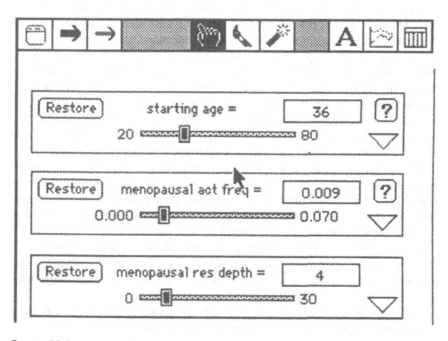

FIGURE 23.2

second graph, only this time set the slider Starting Age to 70 years. In order to answer the questions below, follow the instructions just given to close your second graph, and reopen either graph by double clicking the mouse on the graph to compare simulation results. Unlock and close the graphs when you have answered the questions.

What do the results of this simulation suggest? Would you predict that bone loss is more likely at 70 years than at 20 years? If so, what is your reasoning?

23.4 Examine the Effects of Menopause on Loss of Bone Trabeculae

Now let's assume we are dealing with a woman of postmenopausal age. Change the Starting Age slider to 50. First, let's assume we are dealing with a 50-year-old woman who has not yet entered menopause (estrogen replete). Open graph number one and select Define Graph from the menu Map and I/O. Remove the variables Initial thickness and Minimum Critical Thickness and select the variables Perforated and Resorbed. Click once on OK when finished. Select Run from the menu and count the number of trabeculae lost. Lock and close graph number one. Now, to simulate menopause (estrogen deficient), change the slider Menopausal Act Freq from 0 to 0.07. The simulation assumes an activation frequency, or turnover rate of trabecular bone between 20–25% per year, and we will assume that menopause will increase bone turnover, and at the most will double the activation frequency (0.07). Increasing the activation frequency means that more trabeculae are undergoing remodeling at any point in time, consistent with increased bone turnover observed in menopause. To simulate increased osteoclastic activity, change the slider Menopausal Res Depth from a value of 0 to a value of anywhere above 10. We want to look at the number of trabeculae lost via perforation (overly excessive osteoclastic activity) or full resorption (those trabeculae too thin to survive normal osteoclastic activity). Open graph number two and select Run from the menu. Count the number of trabeculae lost. Lock your graph as before when the simulation is complete by clicking the mouse once on the lock at the bottom left of the graph. Compare the two graphs to answer the questions that follow. After you have answered the questions, Unlock and close the graphs. Figure 23.3 represents one outcome with the Menopausal Resorption Depth set to a value of 15.

Why would altering the variables Menopausal Act Freq and Menopausal Res Depth affect bone loss? Based on your results, what broad generalizations can you make about the effects of estrogen, or the loss of it, on the third lumbar vertebra?

[Note: As the simulation proceeds, one observes a series of vertical lines filling in the graph, each of which corresponds to one remodeling event. The model is removing one trabecular unit from the stock that represents the third lumbar vertebra, and is initiating a remodeling event upon it. The upper line indicates the initial thickness of a single bone trabecula, and the lower

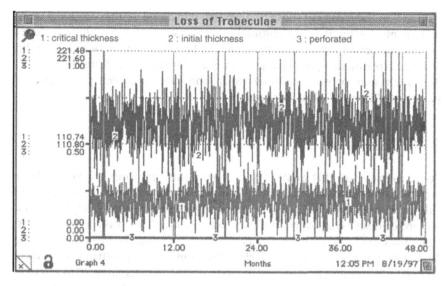

FIGURE 23.3

line indicates the minimum critical thickness below which there is a very high probability that the trabecula will not survive the remodeling process. Any time the upper and lower lines intersect, it is likely that a loss will occur, and this will be shown by the occurrence of a perforation (indicated by a long vertical excursion in the graph of Perforated; see Figure 23.3). At the end of the process, the surviving trabeculae are returned to the stock that represents the third lumbar vertebra.

There are two processes that can lead to loss of bone. One may be likened to perforation, or completely cutting through the bone; the other may be likened to girdling, as when a beaver chews off bark around a tree. The tree may not fall immediately, but chances are very high that it will not survive. The total loss of trabeculae is given by the sum of Perforated and Resorbed.]

23.5 Effects of Drug Treatment on Bone Remodeling

Now let's theorize that one drug interventional treatment inhibits only osteoclastic activity (resorption) while a second interventional treatment inhibits both osteoclastic activity and the increased activation frequency associated with menopause. Change the sliders Menopausal Act Freq and Menopausal Res Depth to correspond to 0.07 and a low value between 0 and 5, respectively. Open graph one and select Run from the menu. Lock graph one when it is finished running. To simulate the second interventional treatment, change Menopausal Act Freq to a value near 0 and leave Menopausal Res Depth alone. Open the second graph and select Run from the menu. Be sure to lock your graph by clicking once on the lock icon on the bottom left of the graph. Compare the two graphs to answer the following questions.

Based on the results of these two graphs, what kind of interventional treatment is more effective at inhibiting bone loss? Why? How would one model effects of calcium supplementation early in life? Similarly, what might be the effects of treatment with a synthetic adrenal glucocorticoid hormone such as dexamethasone or prednisolone? Or an imbalance between production of parathyroid hormone and calcitonin?

23.6 Outcomes

Students who use this model typically respond very enthusiastically. They are able to observe simulated bone loss in a novel and dynamic manner, and this appears to produce a much greater effect than typically results from a lecture or discussion. Moreover, each student is able to control the process. The only drawback we have observed is that networked computers running from a single server perform the simulations much slower than noted when **STELLA®** is running in stand-alone mode on a single machine. This is a case in which some guidance is necessary, because the model is very complex.

References

Delmas, P.D. "Clinical use of biochemical markers of bone remodeling in osteoporosis." *Bone.* 13 (1992): S17–S21.

Hillner, B.E. "Estrogen therapy for geriatric osteoporosis—Just one ball in a complex juggling act." *So. Med. J.* 85 (1992): 2S10–2S16.

Jaros, G.G., T.G. Coleman, and A.C. Guyton. "Model of short-term regulation of calcium-ion concentration." *Simulation.* 32 (1979): 193–204.

Melton, L.J. "Epidemology of osteoporosis." *Bailleres Clin. Obstet. and Gynecol.* 5 (1991): 785–805.

Mosekilde, Li. "Age-related changes in vertebral trabecular bone architecture—assessed by a new method." *Bone.* 9 (1988): 247–250.

Mosekilde, L. "Osteoporosis and calcium." *J. Int. Med.* 231 (1992): 145–149.

Mosekilde, Li. and Le. Mosekilde. "Sex differences in age-related changes in vertebral body size, density, and biomechanical competence in normal individuals." *Bone.* 11 (1990): 67–73.

Mosekilde, Li., Le. Mosekilde, and C.C. Danielson. "Biomechanical competence of vertebral trabecular bone in relation to ash density and age in normal individuals." *Bone.* 8 (1987): 79–85.

Mosekilde, Li., S.E. Weisbrode, J.A. Safron, H.F. Stills, M.L. Jankowsky, D.C. Ebert, C.C. Danielson, C.H. Sogaard, A.F. Franks, M.L. Stevens, C.L. Paddock, and R.W. Boyce. "Calcium-restricted ovariectomized sinclair s-1 minipigs: an animal model of osteopenia and trabecular plate perforation." *Bone.* 14 (1993): 379–382.

Smith, K.H. "Computer simulation of bone remodeling and osteoporosis as a tool for medical education." Master's thesis. Athens, Ga: University of Georgia (1995).

Wardlaw, G.M. "Putting osteoporosis in perspective." *J. Am. Diet. Assoc.* 93 (1993): 1000–1006.

Weaver, C.M. "Age-related calcium requirements due to changes in absorption and utilization." *J. Nutr.* 124 (1994): 1418S–1425S.

24

Positive and Negative Feedback: Insulin and the Use of Fatty Acids and Glucose for Energy

> Such cycles may be found everywhere, wherever Systems exist. In Fashion, skirts go up and down, neckties become wider or thinner. In Politics, the mood of the nation swings Left or Right. Sunspots advance or retreat . . . The pragmatic Systems-student neither exhorts nor deplores, but merely notes the waste of energy involved in pushing the wrong way against such trends.
>
> John Gall,
> *Systemantics*, 1988, p. 34.

Every student of physiology learns the concept of negative feedback, whether it be in terms of baroreceptors and blood pressure, the control of glucose metabolism by insulin, calcium regulation by parathyroid hormone, or sodium regulation of renin, angiotensin, and aldosterone. Due to the growing prevalence of diabetes and the pioneering identification of insulin as the major regulator of blood glucose by Banting and Best, many kineticists have developed dynamic models of insulin secretion and glucose dynamics (Bergman, 1989; Biermann, 1994; Gatewood et al., 1970; Segre et al., 1973). Part of the reason for the interest is the idea that the serious conditions that affect patients with diabetes (such as heart and kidney disease, retinal degeneration and blindness, loss of peripheral nerve function, and bouts of gangrene) may be ameliorated with better glycemic control (Bellomo et al., 1982). A highly pragmatic reason for this interest is not only to understand the distinctions among the different types of diabetes, but also to attain better blood glucose control by use of the artificial pancreas for detection of blood glucose and infusion of insulin (Abisser et al., 1974; Hauser et al., 1994).

A model of an oral glucose tolerance test is included in the tutorial materials obtained when **STELLA®** is purchased, and it would be superfluous to duplicate that model. However, in the context of bioenergetics, a related topic is the ability of insulin and the counterregulatory hormones (epinephrine, cortisol, glucagon, and growth hormone) to control the usage of glucose, amino acids, and fatty acids for energy. This is an extremely important topic apart from the subject of diabetes because the supply of glucose and glycogen only lasts for half a day or so, whereas the fat reserve is usually enough to provide energy for a few months. Whereas any tissue can make use of glucose, the red blood cells and most of the neurons in the brain cannot directly utilize fatty acids. An understanding of the give-and-take regulation of macro-nutrient metabolism is an excellent example of what is called Le Chatelier's Principle: Any process sets up conditions that oppose its continued operation. Or, to phrase this more directly, why doesn't a low-fat diet cause a person to lose more fat? Let us link a model of glucose regulation to fatty acid metabolism and seek an answer.

24.1 The Effect of Insulin on Fuel Use

Sturis and colleagues (1991) have described model equations for simulating short-term (ultradian) oscillations in insulin secretion and glucose utilization. The outcome of this regulation is that mixed meals enhance insulin secretion and cause the body to increase glucose utilization while simultaneously inhibiting the use of fat. This causes the Respiratory Quotient (RQ) to shift to values between 0.9–1.0 (glucose use) or even above 1.0 (fat synthesis). After the postprandial period, insulin secretion subsides, and the non-protein RQ declines toward fasting levels. It can decrease to values as low as 0.7, indicating almost exclusive use of fat (see Figure 24.1). The use of the RQ in this manner subtracts out any use of amino acids for energy, which normally is about 12–15% of the total. However, under fasting conditions, there is an obligatory production of at least 180 grams of glucose per per day from noncarbohydrate sources that include glycerol (released from triacyl-

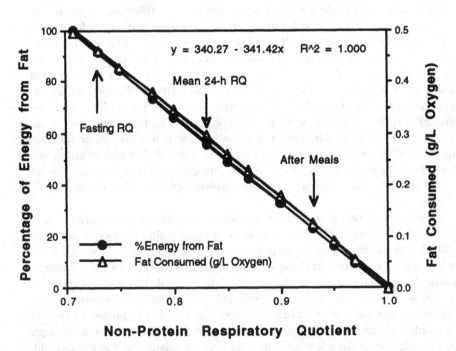

FIGURE 24.1

glycerol during lipolysis), lactic acid (produced during muscle metabolism), and amino acids such as alanine and aspartic acid.

24.2 A Model of Glucose and Fat Regulation

A dynamic model of insulin and glucose regulation was generated and modified in order to demonstate the effect of meals on the relative use of glucose and fatty acids. The model is comprised of three sectors as shown in Figure 24.2A, B, and C.

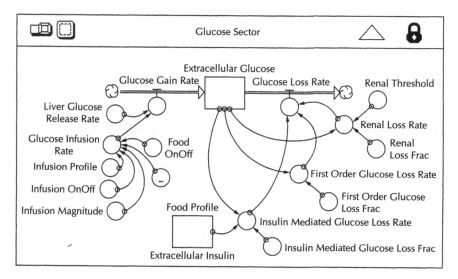

FIGURE 24.2A

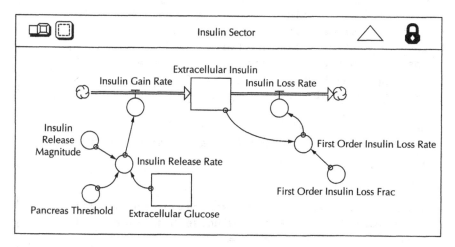

FIGURE 24.2B

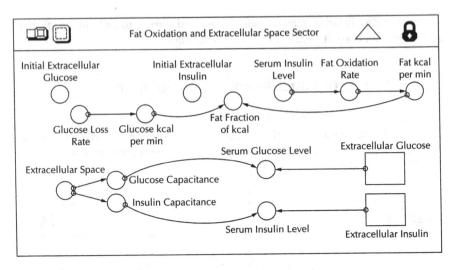

Figure 24.2C

Equations for the Model of Insulin and Substrate Usage

```
Fat Oxidation and Extracellular Space Sector:
Extracellular_Space = 15000 {ml}
Fat_Fraction_of_kcal =
Fat_kcal_per_min/(Fat_kcal_per_min +
Glucose_kcal_per_min)
Fat_kcal_per_min = Fat_Oxidation_Rate * 9
Fat_Oxidation_Rate = 0.3 * (10 - Serum_Insulin_Level)/10
Glucose_Capacitance = Extracellular_Space/100 {dl}
Glucose_kcal_per_min = Glucose_Loss_Rate * 0.11
Initial_Extracellular_Glucose = 12{g}
Initial_Extracellular_Insulin = 1.0 {U}
Insulin_Capacitance = Extracellular_Space/100 {dl}
Serum_Glucose_Level =
Extracellular_Glucose/Glucose_Capacitance * 1000
{mg/dl}
Serum_Insulin_Level =
Extracellular_Insulin/Insulin_Capacitance * 1000
{mg/dl}

Glucose Sector:
Extracellular_Glucose(t) = Extracellular_Glucose(t -
dt) + (Glucose_Gain_Rate - Glucose_Loss_Rate) * dt
INIT Extracellular_Glucose =
Initial_Extracellular_Glucose
```

```
INFLOWS:
Glucose_Gain_Rate = Liver_Glucose_Release_Rate +
Glucose_Infusion_Rate
OUTFLOWS:
Glucose_Loss_Rate = First_Order_Glucose_Loss_Rate +
Insulin_Mediated_Glucose_Loss_Rate + Renal_Loss_Rate
First_Order_Glucose_Loss_Frac = 0.17 {/hr}
First_Order_Glucose_Loss_Rate = Extracellular_Glucose *
First_Order_Glucose_Loss_Frac
Food_OnOff = 1 {1 = On, 0 = Off}
Glucose_Infusion_Rate = Infusion_Magnitude *
(Infusion_OnOff * Infusion_Profile + Food_Profile *
Food_OnOff)
Infusion_Magnitude = 80 {gm/hr}
Infusion_OnOff = 0 {1 = On, 0 = Off}
Infusion_Profile = STEP(1,0) - STEP(1,1)
Insulin_Mediated_Glucose_Loss_Frac = 0.64 {/hr/U}
Insulin_Mediated_Glucose_Loss_Rate =
Insulin_Mediated_Glucose_Loss_Frac *
Extracellular_Insulin * Extracellular_Glucose
Liver_Glucose_Release_Rate = 8.4 {g/hr}
Renal_Loss_Frac = 0.5 {/hr}
Renal_Loss_Rate = IF (Extracellular_Glucose >
Renal_Threshold) THEN (Extracellular_Glucose -
Renal_Threshold) * Renal_Loss_Frac ELSE 0
Renal_Threshold = 37.5 {g}
Food_Profile = GRAPH(MOD(TIME,24))
(0.00, 0.00), (1.00, 0.001), (2.00, 0.001), (3.00,
0.059), (4.00, 0.06), (5.00, 0.059), (6.00, 0.025),
(7.00, 0.002), (8.00, 0.022), (9.00, 0.079), (10.0,
0.08), (11.0, 0.078), (12.0, 0.058), (13.0, 0.02),
(14.0, 0.019), (15.0, 0.059), (16.0, 0.101), (17.0,
0.106), (18.0, 0.098), (19.0, 0.037), (20.0, 0.012),
(21.0, 0.006), (22.0, 0.002), (23.0, 0.001), (24.0,
0.002)


Insulin Sector:
Extracellular_Insulin(t) = Extracellular_Insulin(t -
dt) + (Insulin_Gain_Rate - Insulin_Loss_Rate) * dt
INIT Extracellular_Insulin =
Initial_Extracellular_Insulin
INFLOWS:
Insulin_Gain_Rate = Insulin_Release_Rate
OUTFLOWS:
Insulin_Loss_Rate = First_Order_Insulin_Loss_Rate
```

```
First_Order_Insulin_Loss_Frac = 0.5 {/hr}
First_Order_Insulin_Loss_Rate = Extracellular_Insulin *
First_Order_Insulin_Loss_Frac
Insulin_Release_Magnitude = 0.092 {/hr/g}
Insulin_Release_Rate = IF (Extracellular_Glucose >
Pancreas_Threshold) THEN Insulin_Release_Magnitude *
(Extracellular_Glucose - Pancreas_Threshold) ELSE 0
Pancreas_Threshold = 7.5 {g}
```

Running the model produces a daily cycle in which blood glucose increases after meals, and stimulates insulin secretion. The insulin enhances glucose utilization and storage, eventually causing blood glucose to fall. At the same time, use of fatty acids for fuel is suppressed, as indicated in the Figure 24.3. As the energy in the last meal is utilized, the blood glucose drops and use of fatty acids increases due to a gradually increased rate of lipolysis (cleavage of tryacylglycerol to yield free fatty acids and glycerol). This causes reciprocal use of these two fuels, which together usually account for 80–90% of our daily energy needs.

A feature of this model that was not used elsewhere in this book was the Modulus function, which creates a 24-h clock or timing device in the form of MOD(TIME,24). This was used to generate a food profile in the Glucose Sector. The model is constructed so that the food profile may be switched on or off using a toggle switch (1 = On, 0 = Off), and one may substitute a glucose infusion such as would be done in an oral glucose tolerance test.

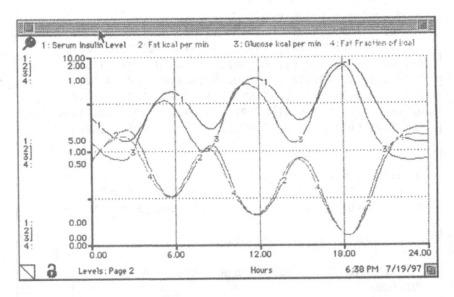

FIGURE 24.3

The student is encouraged to refer to the clinical literature concerning models of glucose regulation and consider models of insulin-dependent diabetes mellitus (IDDM) and non-insulin-dependent diabetes mellitus (NIDDM). Worldwide, the prevalence of NIDDM has increased significantly, and the disorder is now being seen among populations such as those in Asia where there was virtually no record of it before the twentieth century. Models of NIDDM differ from those of IDDM in that the primary defect is thought to be lack of sensitivity of the insulin receptor as the individual becomes obese. Models of this problem are presented by Biermann (1994) and by Quon and Campfield (1991).

References

Abisser, A.M., B.S. Leibel, T.G. Ewart, Z. Dadovac, C.K. Botz, W. Zingg, H. Schipper, and R. Gander. "Clinical control of diabetes by the artificial pancreas." *Diabetes* 23 (1974): 397–404.

Bellomo, J., P. Bruneti, G. Calabrese, D. Mazotti, E. Sarti, and A. Vincenzi. "Optimal feedback glycaemia regulation in diabetics." *Med. Biol. Eng. Comp.* 20 (1982): 329–335.

Bergman, R.N. "Toward physiological understanding of glucose tolerance. Minimal model approach." *Diabetes* 38 (1989): 11512–1527.

Biermann, E. "Simulation of metabolic abonormalities of Type II diabetes mellitus by use of a personal computer." *Comp. Meth. Progr. Biomed.* 41 (1984): 217–229.

Edelstein-Keshet, L. *Mathematical Models in Biology.* New York: Random House/Birkhauser Mathematics Series, 1988.

Gatewood, L.C., E.L. Ackerman, J.W. Rosevear, and G. Molnar, "Modeling blood glucose dynamics." *Behav. Sci.* 15 (1970): 72–87.

Hauser, T., L. Campbell, E. Kraegen, and D. Chisholm. "Glycemic response to an insulin dose change: computer simulator predictions vs. mean patient responses." *Diabetes Nutr. Metab.* 7 (1994): 89–95.

Quon, M., and L. Campfield. "A mathematical model and computer simulation study of insulin receptor regulation." *J. Theor. Biol.* 150 (1991): 59–72.

Segre, G., G.L. Turco, and G. Vercellone. "Modeling blood glucose and insulin kinetics in normal, diabetic, and obese subjects." *Diabetes* 22 (1973): 94–103.

Sturis, J., K.S. Polonsky, E. Mosekilde, and E. van Cauter. "Computer model for mechanisms underlying ultradian oscillations of insulin and glucose." *Am. J. Physiol.* 260 (1991): E801–E809.

25

A Multistage Model for Tumor Progression

The universe is not actually malignant, it only seems so.

John Gall,
Systemantics, p. 6

Because cancer is among the leading causes of death in the United States and most developed societies, it is of great medical and scientific interest. In the clinic, one deals with screening, diagnosis, and treatment; however, the individual who has not developed the disease is most interested in risk reduction. In either case, knowledge about the ways that cancer can develop is important. Just as the research literature concerning cancer is immense, so too are there many different kinds of mathematical models of cancer. Among those of interest are models of cell cycle kinetics (Kimmel and Axelrod, 1991; Novak and Tyson, 1993), carcinogenicity (Ioannides et al., 1993), models based on tumor growth (Mehl, 1991), multistage models of tumor development (Moolgavkar and Knudson, 1981; Stein and Stein, 1990; Moolgavkar and Luebeck, 1992), and metastasis (spreading of a cancer from its site of origin; Tracqui, 1995), and models of treatment outcomes (Retsky et al., 1994; Duchting et al., 1996). Growth models have already been discussed. Here let us consider a multistage model of tumor development, because most tumors require mutations in three or more genes involved in growth control (cellular protooncogenes). The model used here is based on colon cancer, for this cancer affects men and women alike, and is known to be influenced by dietary and lifestyle-related factors.

25.1 Molecular Biology and Staging of Colorectal Cancer

Cells in the mucosa of the large bowel are exposed to many noxious substances and are replaced at a high rate by cell division that occurs in the intestinal crypts. The cells then move up the walls of the crypts to the surface of the mucosa, from which they eventually dislodge and are sloughed off, to meet their demise in the stool. Notwithstanding this wretched fate, the progenitor cells divide frequently, which exposes them to the danger of mutation. This takes place because mutations occur most often and may be fixed in cells that divide frequently, whereas in inactive cells there is much less like-

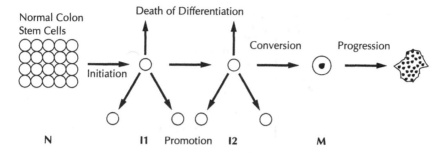

FIGURE 25.1. Basis for a mathematical model of tumorigenesis (Moolgavkar and Knudson, 1981). Growth properties of normal colonic epithelium (N) are altered by mutations that convert it into abnormally growing intermediate stages (I1 and I2), followed by metastasis (M), or changes that cause loss of association with the normal epithelium. Compare with the genetic model shown in Figure 25.2.

lihood that a mutation will occur or be passed on to a daughter cell. A diagram of the process is shown in Figure 25.1.

In describing the process of carcinogenesis, it is clear that accumulation of specific genetic mutations is an essential and obligatory feature. Agents that directly damage the DNA, such as chemical intercalating agents, ionizing radiation, and free-radical generators are called *tumor initiators* (Fig. 25.1). Modifying agents such as dietary fats and fiber are thought to affect the process of *tumor promotion* by which the mutations are fixed and the number of cells possessing the mutation increases. Tumor progression then involves acquisition and fixing of further mutations. These genetic processes are thought to be causally related to the phenotypic stages of cancer that are classified by the pathologist or oncologist (Allan and Sacks, 1995). In the early stages of tumor development, it is very likely that the acquired mutations are not random, but must affect specific genes involved in growth control. *Proto-oncogenes* are genes involved in normal control of growth, and mutations that cause abnormal regulation of these genes or changes in the proteins they generate may cause elevated rates of cell proliferation. The *tumor suppressor genes*, in contrast, normally function to inhibit growth or cell division. Mutation of tumor suppressor genes therefore permits cells to replicate in an uncontrolled manner.

Colorectal cancer has been classified into three major types (Frazier, 1995). These include (1) a "sporadic" form in which there is no clear family history of the disease; (2) inherited adenomatous polyposis coli, which includes familial adenomatous polyposis and Gardner's syndrome; and (3) hereditary nonpolyposis colon cancer. The heritable forms of colorectal cancer have provided evidence of genes whose mutations predispose to early development of cancer, but the multistage model is intended to describe the occurrence of noninherited forms of cancer.

Morphologically, colon cancer develops through a series of stages in which

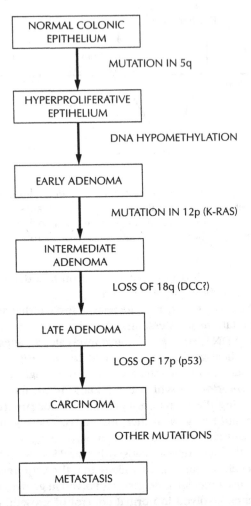

FIGURE 25.2. Molecular events that correlate with tumor progression (Fearon and Vogelstein, 1990). The stages of cancer are noted in the boxes, and the chromosomal or mutational changes associated with each stage are indicated to the right of the arrows.

there is increasing loss of cell growth control and increasingly abnormal cytology; these changes are called *dysplasia*. The multistage model accounts for the concept that a single mutation may cause abnormal growth, but that alone is almost never sufficient to cause a tumor to develop. Instead, a mutation transforms a normal colon cell into an intermediate cell, one that has acquired a mutation and will now pass it on during each cell division. The tumor progresses from its beginnings in normal colonic epithelium to foci in which the epithelium has lost growth control and has become hyperplastic (Fig. 25.2). Cells in a hyperplastic epithelium exit the quiescent phase (G_0) of the cell cycle and enter cell division (DNA synthesis or S Phase and mitosis) more frequently than cells in the normal epithelium. This change is associated with a mutation in chromosome 5q. It has also been observed that DNA in hyperproliferative epithelia is hypomethylated, and this loss of methylation causes the chromosomes to condense improperly (Fearon and Vogelstein, 1990). This change increases the incidence of other mutations due to chromosomal loss or rearrangement.

In general, foci of hyperproliferative epithelium result from clonal proliferation of individual cells that have acquired a mutation, rather than from multiple mutations in many cells. Overgrowth of cells may lead to formation of polyps, which are outgrowths in the colonic mucosa; these do not necessarily have malignant potential. However, polyps with abnormal cytology and those of larger sizes have greater potential for malignancy. The transformation of apparently benign polyps into adenomas of increasing malignancy depends upon mutations in the protooncogene, k-*ras*, and loss of a gene called *dcc* (deleted in colorectal carcinomas; Cho and Vogelstein, 1992). Development of the more malignant carcinoma is associated with loss of a tumor suppressor gene called p53, which normally functions to prevent progression of the cell cycle. Several other mutations are associated with colon cancer, but analysis suggests that most cancers have acquired a minimum of 4–5 mutations (Fearon and Vogelstein, 1990).

25.2 A Multistage Model for Colon Cancer

If a cell in the first population of intermediate cells accumulates a second mutation, it will be further transformed into a second population of intermediate cells, as depicted in Figure 25.3 in a model developed by Moolgavkar and

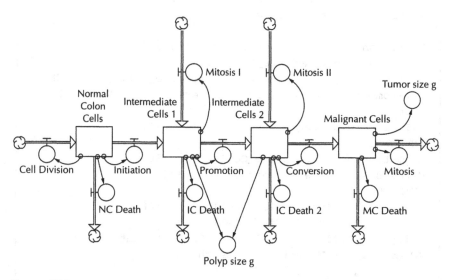

FIGURE 25.3

Luebeck (1992). All populations of intermediate cells have the potential to contribute mass to one or more polyps, which are regarded as initially benign outgrowths created by nature solely to maintain humility among powerful leaders whose endoscopic results are reported annually in the press. However, a third mutation must be acquired to achieve loss of growth control and progression into the population of malignant cells. The malignant cells are doomed to divide exponentially in place until a mutation is acquired that causes loss of cell-cell or cell-matrix contact. Thereafter, the tumor cells invade the intestinal lining and seek a more nutritious venue elsewhere, in the liver perhaps, if not the bone marrow, usually with catastrophic results for their unwilling host.

The diagram in Figure 25.3 represents the model of Moolgavkar and Luebeck (1981) converted into a **STELLA®** model. This model identifies the stages of initiation, promotion, and conversion to malignancy. If it were to be modified to demonstrate the entire series of stages identified by Fearon and Vogelstein (1990), it would be necessary to add more intermediate stages, but the simpler model conveys the idea of different levels of loss of growth control.

Equations for the Model of Colon Cancer

```
Intermediate_Cells_1(t) = Intermediate_Cells_1(t - dt)
+ (Initiation + Mitosis_I -Promotion - IC_Death) * dt
INIT Intermediate_Cells_1 = 1
INFLOWS:
Initiation = Normal_Colon_Cells * 0.107
Mitosis_I = Intermediate_Cells_1 * 2
OUTFLOWS:
Promotion = Intermediate_Cells_1 * 0.307
IC_Death = Intermediate_Cells_1 * 1
Intermediate_Cells_2(t) = Intermediate_Cells_2(t - dt)
+ (Promotion + Mitosis_II - Conversion - IC_Death_2)
* dt
INIT Intermediate_Cells_2 = 1
INFLOWS:
Promotion = Intermediate_Cells_1 * 0.307
Mitosis_II = Intermediate_Cells_2 * 2
OUTFLOWS:
Conversion = Intermediate_Cells_2 * .100
IC_Death_2 = Intermediate_Cells_2 * 2
Malignant_Cells(t) = Malignant_Cells(t - dt) +
(Conversion + Mitosis - MC_Death) * dt
INIT Malignant_Cells = 0
```

```
INFLOWS:
Conversion = Intermediate_Cells_2 * .100
Mitosis = Malignant_Cells * 6
OUTFLOWS:
MC_Death = Malignant_Cells * 4
Normal_Colon_Cells(t) = Normal_Colon_Cells(t - dt) +
(Cell_Division - Initiation - NC_Death) * dt
INIT Normal_Colon_Cells = 10^8
INFLOWS:
Cell_Division = Normal_Colon_Cells * 100
OUTFLOWS:
Initiation = Normal_Colon_Cells * 0.107
NC_Death = Normal_Colon_Cells * 99.893
Age = 1
Diet = 0
DNA_Mutations = (Age + Diet + Heredity + Other) * 0
Heredity = 0
Other = 0
Polyp_size_g = (Intermediate_Cells_1 +
Intermediate_Cells_2)/10^8
Tumor_size_g = Malignant_Cells/10^8
```

25.3 Using the Model of Tumor Growth

This model is among the simplest presented in this book. Each population of cells is characterized by a "birth rate" that depends upon the rate at which mutations accrue in the prior cell population and the intrinsic rate of cell division. This model is designed to simulate changes over several years, so the rate parameters are given using a time frame of years. Mammalian cells in culture can divide with half-times of about 24 h, but if you try this value in the model (0.693 d^{-1} is about 253^{-1} y!), you will find that this is plainly not the rate at which tumor cells divide, despite their fearsome reputation. Each population is also depleted by two events: Either cells may acquire mutations and move into the next population, or they may die. Death is most commonly by sloughing off into the stool, but also may be by apoptosis (programmed cell death). Whether loss by sloughing is the same as necrosis (the other usual fate of cells) is not clear, nor does it make a difference in the model.

Recall that tissues may contain 10^8 cells per gram. We shall have to plot the data in semilogarithmic form just to keep track of the cells. In Figure 25.4, observe that even our intermediate cell populations do not reach a value of 10^8 cells until after 7–8 y. This would only represent the equivalent of a gram of tissue, barely palpable in a digital rectal exam! Note also that, as expected, there will be a delay between the time that the first intermediate cell popu-

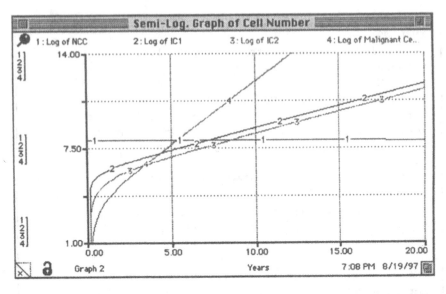

FIGURE 25.4

lation begins to accumulate and the time when the second and third populations accumulate. Unfortunately for this particular subject, a mutation was acquired after just two years of our study. The prognosis is poor if the cancer is
not detected early, for 10^{11} cells would amount to a kilogram, no longer likely
to be localized to the colon.

In this context, the idea of risk reduction becomes quite clear; one would
like to prevent the accumulation of multiple mutations. It is just that one is
not certain how to achieve this worthy aim. How does one prevent mutations
in a population of cells lodged in the deepest recesses of the lower gastrointestinal tract? A variety of advice is available from the American Cancer Society (1996). Among the suggestions are consumption of adequate levels of
dietary fiber (about 35 g per day), low intake of saturated fat, generous intakes of cruciferous vegetables (broccoli, for instance), and an aspirin a day.
Given the prevalence of oncogenes, one might add that judicious choice of
one's parents could be desirable!

25.4 Modifications in the Model

The model shown is extremely simple and it would be worthwhile considering various ways to simplify or extend it. For example, mutations are random
events; would it be worthwhile to introduce a mutation rate that is stochastic? In order to fix a mutation, the cell must divide before it dies, and for the
tumor to progress, additional mutations must occur within the hyperproliferative cell population. Is it feasible to model the effects of dietary agents that
are thought to modify the rate of tumor promotion? What is the best way to

model the effect of tumor suppressor genes? In this case, is it necessary to lose both copies of the gene to have an effect, or is there some loss of growth control if only one of the two chromosomes is modified? If the multistage model is applicable to an individual, could it be extended to a population? The student is encouraged to read papers from the References and consider possible changes.

References

Allan, S.M., and N.P.M. Sacks. "Colorectal cancer." In: Horwich, A., ed. *Oncology. A Multidisciplinary Textbook*. London: Chapman and Hall, 1995, 513-528.

American Cancer Society 1996 Advisory Committee on Diet, Nutrition, and Cancer Prevention. Guidelines on diet, nutrition, and cancer prevention—reducing the risk of cancer with healthy food choices and physical activity. *CA-Cancer J. Clin.* 46 (1996): 325-341.

Cho, K.R., and B. Vogelstein. "Genetic alterations in the adenoma-carcinoma sequence." *Cancer* 70 (Suppl. 6, 1992): 1727-1731.

Duchting, W., W. Ulmer, and T. Ginsberg. "Cancer: a challenge for control theory and computer modelling." *Eur. J. Cancer* 32A (1996): 1283-1292.

Fearon, E.R., and B. Vogelstein. "A genetic model for colorectal tumorigenesis." *Cell* 61 (1990): 759-767.

Frazier, M. L. "Molecular basis of colon cancer." In: Freireich, E.J., and S.A. Stass, eds. *Molecular Basis of Oncology*. Oxford: Blackwell Science Publishers, 1995, 317-339.

Ioannides, C., D.F. Lewis, and D.V. Parke. "Computer modelling in predicting carcinogenicity." *Eur. J. Cancer Prev.* 2 (1993): 275-282.

Kimmel, M., and D.E. Axelrod. "Unequal cell division, growth regulation, and colony size of mammalian cells: a mathematical model and anlysis of experimental data." *J. Theoret. Biol.* 153 (1991): 157-180.

Mehl, L.E. "A mathematical computer simulation model for the development of colonic polyps and colon cancer." *J. Surg. Oncol.* 47: 243-252; 1991.

Moolgavkar, S.H., and A.G. Knudson. "Mutation and cancer: a model for human carcinogenesis." *J. Nat. Cancer Inst.* 66 (1981): 1037-1052.

Moolgavkar, S.H., and E.G. Luebeck. "Multistage carcinogenesis: population-based model for colon cancer." *J. Natl. Cancer Inst.* 84 (1992): 610-618.

Novak, B., and J.J. Tyson. "Modeling the cell division cycle: M-phase trigger, oscillations, and size control." *J. Theor. Biol.* 165 (1993): 101-134.

Retsky, M.W., D.E. Swartzendruber, P.D. Barne, and R.H. Wardwell. "Computer model challenges breast cancer treatment strategy." *Cancer Invest.* 12 (1994): 559-567.

Stein, W.D., and A.D. Stein. "Testing and characterizing the two-stage model of carcinogenesis for a wide range of human cancers." *J. Theoret. Biol.* 145 (1990): 95-122.

Tracqui, P. "From passive diffusion to active cellular migration in mathematical models of tumor invasion." *Acta Biotheoretica* 43 (1995): 443-464.

26

The Biokinetic Database

Time, which requires the utmost purity of consciousness to be properly
apprehended, is the most rational element of life . . .

Vladimir Nabokov,
Ada or Ardor, A Family Chronicle

26.1 Epilogue

It is almost axiomatic that time is the least appreciated dimension in biology.
We can grasp at the idea that passage of time is relative, and not the same
thing to a shrew as to a sloth, a mouse, or a man. This relativity of duration
may be related to the rate at which electrons derived from foodstuffs are utilized in our mitochondria, such that one marker of time's passing is irrepara-
ble oxidative damage to the genetic code in our somatic cells, a process to
which we are insensible. Even admitting that time is relative, it seems that
there may be evolutionary advantages to experiencing it quickly, and other
advantages (and drawbacks) to perceiving the passage of time slowly. For this
reason alone, the dimension should be of interest to the biologist, and all the
more when one develops an interest in biological rhythms and the process of
aging, however nebulous it may seem.

We need a tool to investigate time and time-dependent processes. Therefore, one must begin to grapple with rates of change, and how a system must
be designed so that all the different rates that apply to its parts are somehow
integrated, for so they must be. There is hope that the personal computer can
be our servant in thinking about the temporal dimension, for it at least allows
one to contemplate rates, to integrate different sorts of quantitative data, and
to begin thinking about system design.

To restate a second theme: The personal computer offers capabilities that
could enhance anyone's quantitative understanding, from high school student to senior research scientist, regardless of background. Programs for dynamic modeling offer many advantages for use in education, and should be
employed to a much higher degree than they are now. Algebra is the only
prior training that students need to make excellent use of the programs. Yes,
training in the calculus can help, but in many ways, the new tool offers the
capabilities of calculus without requiring so much theory. **STELLA**® and related programs can help students understand the point of calculus, for they
allow equations to be solved and provide a window to view the outcomes.

Why not harness these immense capabilities for training in the quantitative disciplines of biomedical science?

Many very talented scientists have attempted to summarize ways in which the human or animal system responds to nutrients, hormones, and drugs by reporting the dynamics of our cellular and molecular constituents using isotopic tracers. Unfortunately, it is with difficulty that the average biologist understands messages conveyed in terms of quantitative relationships. Learning to use the personal computer cannot convert a person who lacks specific training into a kineticist, but it can enable the intelligent student to understand what is meant by mathematical formulas, the shape of an accumulation curve, or the ways in which control points can be (and are) employed to regulate the human system.

Software for simulation allows one to approach the concept of system dynamics. It provides a tool that will allow a person to do what is otherwise very difficult, if not impossible, and can even make the work fun and adventuresome. Some maintain that this discipline is not for the naive or the untrained, but one is reminded of the well-intentioned advice offered to Gregor Mendel by a renowned nineteenth century botanist: that he should give up the study of inheritance in garden peas, and focus on an interesting plant— hawkweed, for instance! However, whether or not the personal computer aids in more great discoveries, or even personal revelations, it at least can be used to interpret data that already exist.

The only drawback to this approach is the one shared by all software purchases. Software costs as much as textbooks, and if used on a networked system, a way must be found to obtain licensing fees and periodic upgrades. However, a Run-Time version of **STELLA**® has just been posted on the Internet, and one may anticipate this will be the standard of the future, so that students may at least try the software without purchasing it. Run-Time software does not allow changes in models to be saved, so this is only a partial solution, but one must realize that the programs are commercial products and the vendors maintain businesses that deserve our support to the extent that the products meet specific needs. The need for more emphasis on quantitative methods in science education is familiar to all educators.

26.2 Software for Modeling and Simulation

Computer-assisted mathematical modeling is applied in every scientific discipline, particularly in the fields of engineering, computer chip design, and process control. In every area, individuals and research teams have developed software packages appropriate for their specific needs. Moreover, as computer capabilities have escalated, the software has evolved to make use of greater processing speed, larger amounts of memory, and newer interfaces. As in every computer-related field, there are many different commercial vendors and products with many capabilities. Hundreds of different simulation programs exist, and there is a great amount of competition and winnowing. The Society for Computer Simulation publishes an annual directory of simulation software that lists several hundred different varieties of software, a brief description of each, its cost, telephone numbers, and e-mail addresses.

One of the first general modeling programs designed for the biomedical sciences is called Simulation, Analysis and Modeling (SAAM), which was developed at the National Institutes of Health by Mones Berman and colleagues. The program evolved expressly to extract all the possible data from kinetic studies, including numerical values that provide the best fit between the data and specific conceptual models (reviewed by Zech et al., 1991). A modified version called Conversational SAAM (CONSAM) is more user friendly and has been widely used by biomedical scientists, especially for studies involving isotopic tracers (Boston et al., 1981). A new program called SAAM II has been created by Dr. David Foster and colleagues to include the capabilities of SAAM and the ease of use of a menu-driven, graphical user interface. This is also more user friendly, and harnesses contemporary methods of setting up and solving differential equations, performing numerical analysis, and making use of statistics. The SAAM programs may be appropriate for any graduate student or professional scientist who must perform statistical tests and obtain best fits of data with theory. However, it is certainly true that the more sophisticated programs require more time and dedication to learn. Several programs are available that satisfy all requirements; no one should mistakenly purchase a program that lacks a modern interface unless there is a very good reason for using it. As always, prospective modelers are advised to be sure that their computer systems have adequate memory, a compatible central processor, and the proper interface to use the software that is chosen. It is always helpful to talk to colleagues who understand the discipline and the software capabilities before purchasing commercial software.

26.3 Using Published Models

Every scientific discipline seeks to create models; most also make use of kinetic models. This is certainly true of physiology, biochemistry, nutrition, cellular and molecular biology, genetics, epidemiology, pharmacology, and toxicology. Computational methods that rely on various kinds of tracers are also extremely important in clinical medicine, and provide the basis for diagnostic procedures that involve any manner of image scanning. Drs. Merle Wastney, Raymond C. Boston, and colleagues have begun collecting published kinetic models that employ simulation software (principally SAAM and CONSAM). These comprise a part of the published, biokinetic database. Some of these models are available at an Internet site maintained at Georgetown University.

It is impossible and unnecessary to summarize all the models that are available, but it may be worthwhile to provide a few references and starting points. In evaluating models, one should remember that it is much more informative to begin with simple models, rather than the highly complex ones that contain too much detail to be helpful to one's thinking. "'Tis a gift to be simple," as the traditional song says, or to paraphrase the words of a colleague, "If you want to get a lot out of a model, do not put everything you know into it."

26.4 Hypothesis Testing and Publication

Computer simulation is a powerful, quantitative tool that can allow students, postdoctoral fellows, and established investigators to develop capabilities and perform tests that would otherwise be extremely difficult. Simulation can allow students to do experiments that do not require costly laboratory equipment (as long as personal computers are available), yet engage in active learning. In developing a model, one must think, seek information, assimilate, quantify, balance, and experiment. There is no assurance that a student who has only reviewed a published paper will have become engaged in any but the most cursory manner; yet with simulation, one may generate and test hypotheses.

The use of simulation and modeling may be likened to the process of writing using a word processor. This powerful tool can be used competently or superficially; the use of a word processor does not by itself improve grammar, self-expression, or qualify a person to work as a technical writer or editor. In every discipline, however, there is a place for the amateur and the hobbyist. If one wishes to become a biophysicist or biometrician, it is still necessary to train in rigorous disciplines. One hopes that by learning something about modeling with software such as **STELLA®**, which was designed for educational uses, the student will learn to understand more of the information that is available from the specialists.

Word processing software is useful to the scientist because it helps in preparing reports for publication and for grant proposals. The same statement can be made of software for modeling. Still, there is an impediment in that the traditional method of explaining models in the scientific literature is by use of differential equations. The author has found that models developed with **STELLA®** are often sent to referees who are not accustomed to reviewing finite difference equations, and poor communication has at times impeded publication. A treasured letter I received from the editor of a renowned biochemical journal stated, "The Journal does not publish material of this kind; however, when you find one that does, could you please forward me a copy?" I did, and he kindly thanked me for it.

It is extremely important to understand that finite difference equations are an established alternative to analytical solutions of differential equations. The scientist who wishes to publish a **STELLA®** model may be well advised to state at least some of the relationships in the form of differential equations. For example, Goumaz and colleagues (1991) described a model of mRNA metabolism in terms of differential equations but solved the equations using

STELLA®. This is valid, and the paper was published in a highly respected journal. The prerequisite to publishing results of modeling is still the quality of data used to support the model.

26.5 Divergent Origins of Modeling and System Dynamics

STELLA® and similar programs were developed in support of a discipline that is generally called system dynamics, which is rooted in work done by scientists such as Jay Forrester and colleagues (Forrester, 1969; Hayes, 1993). System dynamics is usually used in the context of macroscopic events and economics, an excellent example being the famous book *The Limits to Growth* (Meadows et al., 1972). In another sense, system dynamics is related to work done in the early days of the computer revolution by individuals such as Claude Shannon and Warren Weaver (1949) on information theory, and Ludwig von Bertalanffy (1968) on general system theory. Certainly Mones Berman and colleagues were aware of these disciplines, but the SAAM program dealt with specific problems in modeling and statistical applications concerning biological kinetics. Now, as simulation programs are developed, people with many different interests and kinds of training are interacting.

It remains to be seen whether a lingua franca or common language will ever develop in these related disciplines. It is clear that individuals trained in system dynamics tend to be less oriented toward statistics and to have less mathematical background. According to Dr. Jay Forrester, methods of simulation and the principles of system dynamics should be introduced in grade school, if not kindergarten! Instead of putting an understanding of mathematics at the forefront, this objective would place the tool of mathematics in the background, and emphasize concepts and ideas. The means of achieving this aim relies on the use of stock-and-flow diagrams to teach a more realistic view of causes and consequences for events that occur daily. In a sense, many of those who have learned this point of view are creative amateurs.

If science is to reclaim a sense of playful adventure, and the joy that only the amateur (from *amare*, to love) can experience, the author would hold that system dynamics represents a key. One must also recognize the hard-won attributes and requirement for rigor that separate the amateur from the professional. Thus the scientific world needs to find ways to include both points of view, to attract the kind of youthful enthusiasts, young and old, who will persevere in pursuing the deep problems of nature.

References

Bertalanffy, Ludwig von. *General System Theory*. New York: Braziller, 1968.

Boston, R.C., P.C. Greif, and M. Berman. "Conversational SAAM—an interative program for kinetic analysis of biological systems." *Comput. Progr. Biomed.* 13 (1981): 111–119.

Forrester, J.W. *Urban Dynamics*. Cambridge, Ma.: M.I.T. Press, 1969.

Goumaz, M.O., H. Schwartz, J.H. Oppenheimer, and C. Mariash. "Kinetic model of the response of precursor and mature rat hepatic mRNA-S14 to thyroid hormone." *Am. J. Physiol.* 266 (1994): E1001-1011.

Hayes, B. "Balanced on a pencil point." *Amer. Scientist 81* (1993): 510-516.

Meadows, D.H., D.L. Meadows, J. Randers, et al., eds. *The Limits to Growth*. New York: Universe Books, 1972.

Shannon, C.E., and W. Weaver. *The Mathematical Theory of Communication*. Urbana: University of Illinois Press; 1949.

Society for Computer Simulation, 4838 Ronson Court, Suite L, P.O. Box 17900, San Diego, Cal. Telephone: (619) 277-3888; e-mail, scs@sdsc.edu.

Zech, L.A., D.J. Rader, and P.C. Greif, "Berman's simulation analysis and modeling." *Adv. Exp. Biol. Med.* 285 (1991): 188-199.

Appendix:
Quick Help Guide
to STELLA® Software Mechanics

1.   System Requirements:

Windows™ version:

Minimum System Requirements
80486 processor (386 enhanced mode)
Microsoft Windows™ 3.1
8 MB RAM
16 MB Hard-disk space required
QuickTime™

Recommended System Requirements
Pentium processor or better
Microsoft Windows™ 95
16 MB RAM
VGA display of at least 256 colors
SoundBlaster or compatible sound card

Macintosh® version:

Minimum System Requirements
68020 processor
System 7.1 or higher
8 MB RAM (4 MB available to the program)
12 MB Hard-disk space required
QuickTime™

Recommended System Requirements
68040 processor or better (including Power PC)
16 MB RAM

2.   Overview of the STELLA Operating Environment (Model Layer)

High-level Map/Model/Equations Arrows

Available in the **STELLA Research** version only.

3.   Drawing an Inflow to a Stock

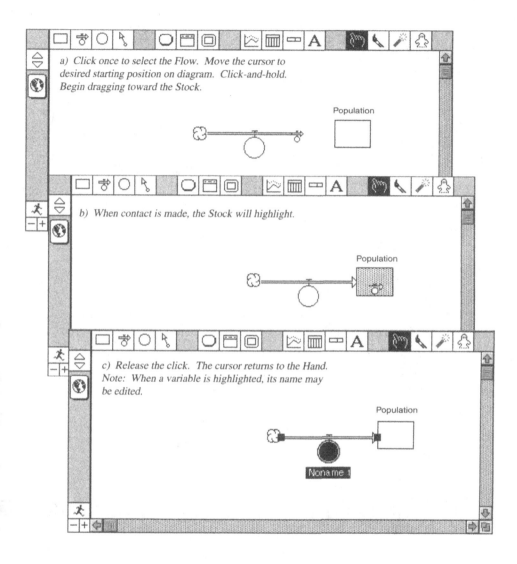

a) Click once to select the Flow. Move the cursor to desired starting position on diagram. Click-and-hold. Begin dragging toward the Stock.

Population

b) When contact is made, the Stock will highlight.

Population

c) Release the click. The cursor returns to the Hand. Note: When a variable is highlighted, its name may be edited.

Population

Noname 1

4.   Drawing an Outflow from a Stock

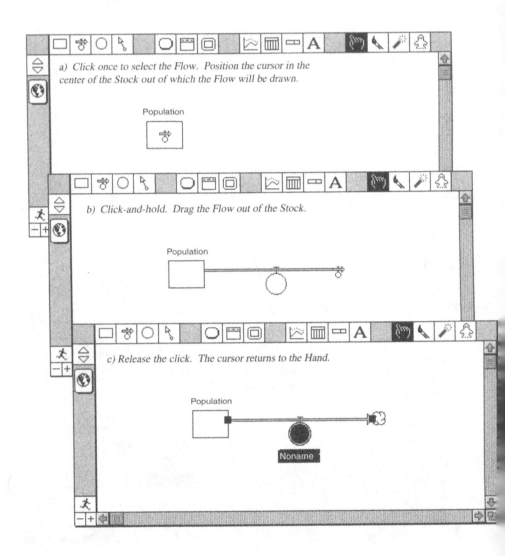

a) Click once to select the Flow. Position the cursor in the center of the Stock out of which the Flow will be drawn.

Population

b) Click-and-hold. Drag the Flow out of the Stock.

Population

c) Release the click. The cursor returns to the Hand.

Population

Noname

5.   Replacing a Cloud with a Stock

(If the Stock is initially not hooked up with the Flow.)

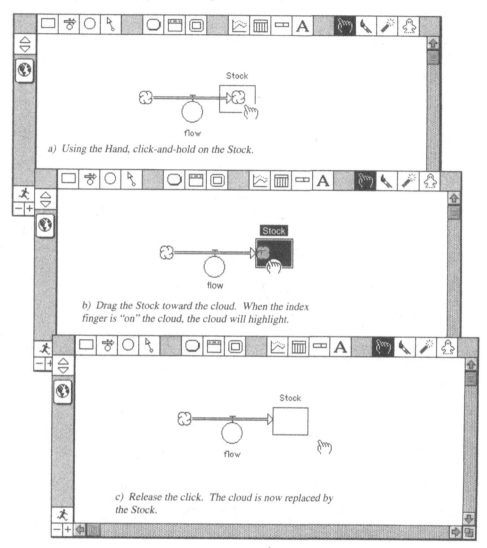

a) *Using the Hand, click-and-hold on the Stock.*

b) *Drag the Stock toward the cloud. When the index finger is "on" the cloud, the cloud will highlight.*

c) *Release the click. The cloud is now replaced by the Stock.*

6.   Bending Flows

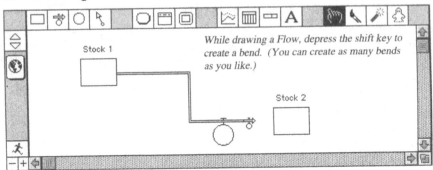

While drawing a Flow, depress the shift key to create a bend. (You can create as many bends as you like.)

Stock 1

Stock 2

7.   Re-positioning Flows

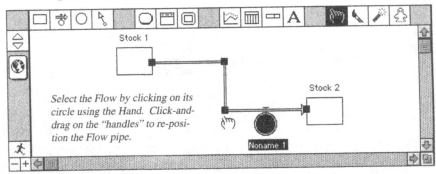

Stock 1

Stock 2

Select the Flow by clicking on its circle using the Hand. Click-and-drag on the "handles" to re-position the Flow pipe.

Noname 1

8.   Reversing Direction of a Flow

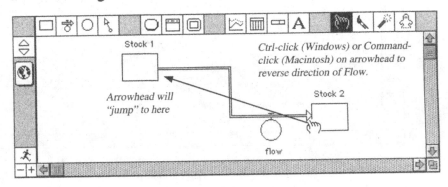

Stock 1

Ctrl-click (Windows) or Command-click (Macintosh) on arrowhead to reverse direction of Flow.

Arrowhead will "jump" to here

Stock 2

flow

9.   Flow Dialog

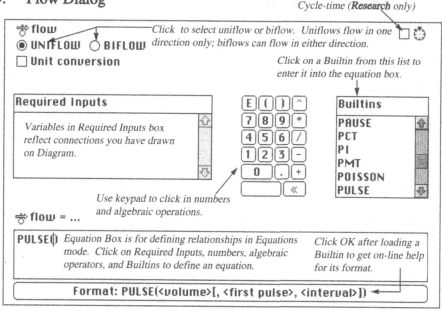

*Cycle-time (**Research** only)*

flow —————— *Click to select uniflow or biflow. Uniflows flow in one*
◉ **UNIFLOW** ○ **BIFLOW** *direction only; biflows can flow in either direction.*

☐ **Unit conversion**

Click on a Builtin from this list to enter it into the equation box.

Required Inputs

Variables in Required Inputs box reflect connections you have drawn on Diagram.

E	(	)	^
7	8	9	*
4	5	6	/
1	2	3	–
0	.	+
	«

Builtins

PAUSE
PCT
PI
PMT
POISSON
PULSE

Use keypad to click in numbers
flow = ...   *and algebraic operations.*

PULSE() *Equation Box is for defining relationships in Equations mode. Click on Required Inputs, numbers, algebraic operators, and Builtins to define an equation.*

Click OK after loading a Builtin to get on-line help for its format.

Format: PULSE(<volume>[, <first pulse>, <interval>]) ◄

10.   Moving Variable Names

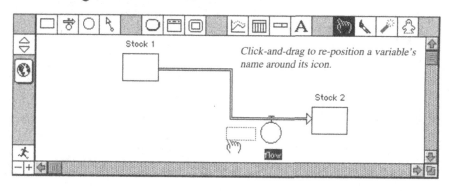

Stock 1

Click-and-drag to re-position a variable's name around its icon.

Stock 2

flow

11. Drawing Connectors

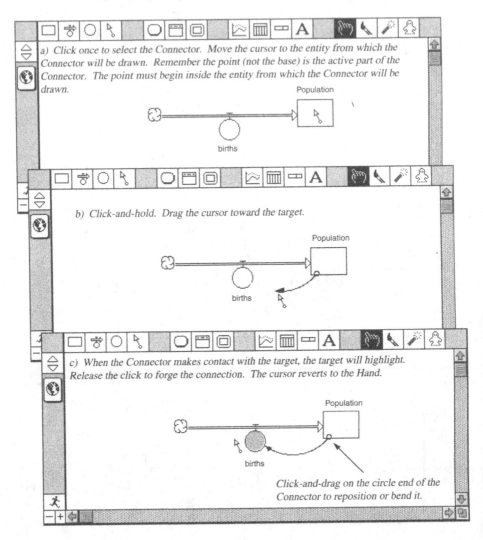

a) *Click once to select the Connector. Move the cursor to the entity from which the Connector will be drawn. Remember the point (not the base) is the active part of the Connector. The point must begin inside the entity from which the Connector will be drawn.*

b) *Click-and-hold. Drag the cursor toward the target.*

c) *When the Connector makes contact with the target, the target will highlight. Release the click to forge the connection. The cursor reverts to the Hand.*

Click-and-drag on the circle end of the Connector to reposition or bend it.

12.   Defining Graphs and Tables

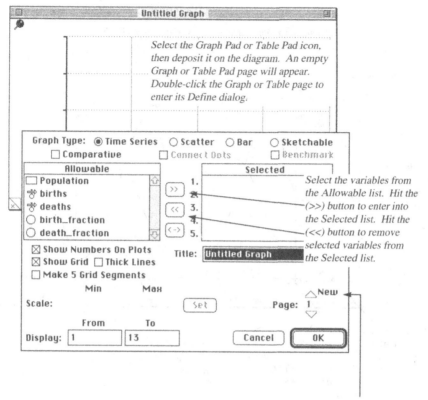

Select the Graph Pad or Table Pad icon, then deposit it on the diagram. An empty Graph or Table Pad page will appear. Double-click the Graph or Table page to enter its Define dialog.

Select the variables from the Allowable list. Hit the (>>) button to enter into the Selected list. Hit the (<<) button to remove selected variables from the Selected list.

Create as many new Graph pages as you like.

13. Other Operations on Graphs and Tables

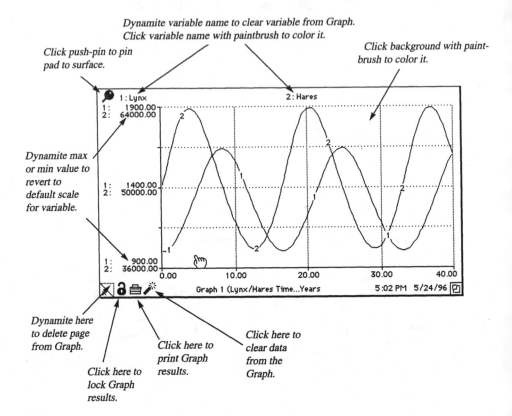

Dynamite variable name to clear variable from Graph.
Click variable name with paintbrush to color it.

Click push-pin to pin
pad to surface.

Click background with paint-
brush to color it.

Dynamite max
or min value to
revert to
default scale
for variable.

Dynamite here
to delete page
from Graph.

Click here to
print Graph
results.

Click here to
clear data
from the
Graph.

Click here to
lock Graph
results.

Bibliography

General References

Anderson, J.G., M. Katzper. *Simulation in the Health Sciences* (Annual Proceedings). San Diego: Society for Computer Simulation, (1996).

Berman, M., S.M. Grundy, B.V. Howard, eds. *Lipoprotein kinetics and modeling*. New York: Academic Press; 1982.

Bertalanffy, L. von. *General System Theory*. New York: Braziller, 1968.

Boston, R.C., P.C. Greif, M. Berman. "Conversational SAAM—an interactive program for kinetic analysis of biological systems." *Comput. Progr. Biomed.* 13 (1981): 111-119.

Brown, D., P. Rothery. *Models in Biology: Mathematics, Statistics, and Computing*. New York: John Wiley, 1993.

Carson, E.R., C. Cobelli, L. Finkelstein. *The Mathematical Modeling of Metabolic and Endocrine Systems. Model Formulation, Identification, and Validation*. New York: John Wiley, 1983.

Coburn, S.P., D.W. Townsend. "Mathematical Modeling in Experimental Nutrition. Vitamins, Proteins, Methods." *Adv. Food Nutr. Res.* 40 (1996): San Diego: Academic Press.

Collins, J.C. "Resources for getting started in modeling." *J. Nutr.* 122 (1992): 695-700.

Department of Health and Human Services. *Biomedical Research Technology Resources: A Research Resources Directory*, 8th ed. Rockville, Md.: Research Resources Information Center; 1990.

Forrester, J.W. *Urban Dynamics*. Cambridge Ma.: M.I.T. Press; 1969.

Green, M.H. "Introduction to modeling." *J. Nutr.* 122 (3 Suppl., 1992): 690-694.

Green, M.H., J.B. Green. "The application of compartmental analysis to research in nutrition." *Annu. Rev. Nutr.* 10 (1990): 41-61.

Hayes, B. "Balanced on a pencil point." *Amer. Scientist* 81 (1993): 510-516.

Katz, J. The determination of mass of metabolites with tracers." *Metabolism* 38 (1989): 728-733.

Meadows, D.H, D.L. Meadows, J. Randers, et al., eds. *The Limits to Growth*. New York: Universe Books; 1972.

Novotny, J.A., L.A. Zech, H.C. Furr, S.R. Dueker, and A.J. Clifford. Mathematical modeling in nutrition: constructing a physiological compartmental model of the dynamics of beta-carotene metabolism." *Adv. Food Nutr. Res.* 40 (1996): 25-54.

Ramberg, C.F., Jr., C.R. Krishnamurti, D. Peter, J.E. Wolff, and R.C. Boston. "Application of models to determination of nutrient requirements: experimental techniques employing tracers." *J. Nutr.* 122, (3 Suppl., 1992) 701-705.

Segel, L.A. *Biological Kinetics*, Cambridge: Cambridge University Press; 1991.

Shannon, C.E., W. Weaver. *The Mathematical Theory of Communication.* Urbana: University of Illinois Press; 1949.

Umpleby, A.M., P.H. Sonksen. "Measurement of the turnover of substrates of carbohydrate and protein metabolism using radioactive isotopes." *Baillieres Clin. Endocrinol. Metab.* 1 (1987): 773-796.

Zech, L.A., D.J. Rader, P.C. Greif. "Berman's simulation analysis and modeling." *Adv. Exp. Biol. Med.* 285 (1991): 188-199.

Biochemical and Physiological Models

Dempsher, D.P., D.S. Gann, R.D. Phair. "A mechanistic model of ACTH-stimulated cortisol secretion." *Am. J. Physiol.* 246 (1884): R587-R596.

Goumaz, M.O., H. Schwartz, J.H. Oppenheimer, C. Mariash. "Kinetic model of the response of precursor and mature rat hepatic mRNA-S14 to thyroid hormone." *Am. J. Physiol.* 266 (1994): E1001-1011.

Kraus, M., P. Lais, B. Wolf. "Systems analysis in cell biology: from the phenomological description towards a computer model of the intracellular signal transduction network." *Experientia* 49 (1993): 245-257.

Mendel, C.M. "The free hormone hypothesis: a physiologically based mathematical model." *Endocrin. Rev.* 10 (1989): 232-274.

Solomon, Y., M.C. Lin, M. Rodbell, M. Berman. "The hepatic adenylate cyclase system. III. A mathematical model for the steady state kinetics of catalysis and nucleotide regulation." *J. Biol. Chem.* 250 (1975): 4253-4260.

Traganos, F., M. Kimmel, C. Bueti, Z. Darzynkiewicz. "Effects of inhibition of RNA or protein synthesis on CHO cell cycle progression." *J. Cell. Physiol.* 133 (1987): 277-287.

Body Composition

Dulloo, A.G., J. Jacquet, L. Girardier. "Autoregulation of body composition during weight recovery in humans: the Minnesota Experiment revisited." *Int. J. Obesity Rel. Metab. Disord.* 20 (1996): 393-405.

Heymsfeld, S.B., M. Waki, J. Kehayias, S. Lichtman, F.A. Dilmanian, Y. Kamen, J. Wang, R.N. Pierson, Jr. "Chemical and elemental analysis of humans in vivo using improved body composition models." *Am. J. Physiol.* 261 (1991): E190-E198.

Cell Cycle and Cell Growth Kinetics

Allman, R., T. Schjerven, E. Boye. "Cell cycle parameters of Escherichia coli K-12." *J. Bacteriol.* 173 (1991): 7970-7974.

Blackett, N.M. "Haemopoietic spleen colony growth: a versatile, parsimonious, predictive model." *Cell Tissue Kinetics* 20 (1987): 393-402.

Heenen, M., V. DeMartelaer, P. Galand. "Psoriasis and cell cycle: a computer simulation." *Acta Derm. Venereol. Suppl. Stockholm* 87 (1987): 73-76.

Prikrylova, D., and D. Bucanek. "Mathematical model of the cell cycle regulation in budding yeasts." *Biomed. Biochim. Acta* 49 (1990): 733-736.

Prothero, J., M. Starling, C. Rosse. "Cell kinetics in the erythroid compartment of guinea pig bone marrow: a model based on ^{3}H-TdR studies." *Cell Tissue Kinetics* 11 (1978): 301-316.

Retsky, M.W., D.E. Swartzendruber, P.D. Bame, R.H. Wardwell. "Computer model challenges breast cancer treatment strategy." *Cancer Invest.* 12 (1994): 559-567.

Sennerstam, R., J.O. Stromberg. "Cell cycle progression: computer simulation of uncoupled subcycles of DNA replication and cell growth." *J. Theor. Biol.* 175 (1995): 177-189.

Tyson, J.J., K.B. Hannsgen. "Cell growth and division: a deterministic, probabilistic model of the cell cycle." *J. Math. Biol.* 23 (1986): 231-246.

Cholesterol

Redgrave, T.G., H.L. Ly, E.C. Quintao, C.F. Ramberg, R.C. Boston. "Clearance of triacylglycerol and cholesteryl ester after intravenous injection of chylomicron-like lipid emulsions in rats and man." *Biochem. J.* 290 (1993): 843-847.

Schwartz, C. C., M. Berman, Z.R. Vlahcevic, L. Swell. "Multicompartmental analysis of cholesterol metabolism in man. Quantitative kinetic evaluation of precursor sources and turnover of high density lipoprotein cholesterol esters." *J. Clin. Invest.* 70 (1982): 863-876.

Stacpoole, P.W., K. Von Bergmann, L.L. Kilgore, L.A. Zech, W.R. Fisher. "Nutritional regulation of cholesterol synthesis and apolipoprotein B kinetics: studies in patients with familiar hypercholeterolemia and normal subjects treated with a high carbohydrate, low fat diet." *J. Lipid Res.* 32 (1991): 2837-48.

Circadian Rhythms

Chandler, W.L. "A kinetic model of the circulatory regulation of tissue plasminogen activator." *Thromb. Haemost.* 66 (1991): 321-328.

Crosthwaite, S.K., J.J. Loros, J.C. Dunlap. "Light induced resetting of a circadian clock is mediated by a rapid increase in frequency transcript." *Cell* 81 (1995): 1003-1012.

Gundel, A., and R.B. Spencer. "A mathematical model of the human circadian system and its applications to jet lag." *Chronobiol. Int.* 9 (1992): 148-159.

Halberg, F., G. Cornelissen, N. Marques, L.M. Barreto, M.D. Marques. "From circadians of the fifties to chronomes in vitro as in vivo." *Arch. Med. Res.* 25 (1994): 287-296.

Kai, M., T. Eto, K. Kondo, Y. Setoguchi, S. Higashi, Y. Maeda, T. Setoguchi. "Synchronous circadian rhythms and activities of cholesterol 7 alpha-hydroxylase in the rabbit and rat." *J. Lipid. Res.* 36 (1995): 367-374.

Klerman, E.B., D.J. Dijk, R.E. Kronauer, C. Czeisler. "Simulations of light effects on the human circadian pacemaker: implications for assessment of intrinsic period." *Am. J. Physiol.* 270 (1996): R271-R282.

Pedersen, M., A. Johnsson. "A study of the singularities in a mathematical model for circadian rhythms." *Biosystems* 33 (1994): 193-201.

Hemodialysis

Thews, O., H. Hutten. "A comprehensive model of the dynamic exchange processes during hemodialysis." *Med. Prog. Technol.* 16 (1990): 145-161.

Folate, Homocysteine, and Heart Disease

von der Porten, A.E., J.F. Gregory, J.P. Toth, J.J. Cerda, S.H. Curry, L.B. Bailey. "In vivo folate kinetics during chronic supplementation of human subjects with deuterium-labeled folic acid." *J. Nutr.* 122 (1992): 1293-1299.

Ubbink, J.B., P.J. Becker, W.J. Vermaak, R. Delport. "Results of B-vitamin supplementation study used in a prediction model to define a reference range for plasma homocysteine." *Clin. Chem.* 41 (1995): 1033-1037.

Protein and Amino Acids

Biolo, G., D. Chinkes, X.J. Zhang, R.R. Wolfe. (Harry M. Vars Research Award) "A new model to determine in vivo the relationship between amino acid transmembrane transport and protein kinetics in muscle." *JPEN-J. Parenter. Enteral. Nutr.* 16 (1992): 305-315.

Cheng, R.N., F. Dworzak, G.C. Ford, M.J. Rennie, D. Halliday. "Direct determination of leucine metabolism and protein breakdown in humans using L-[1-^{13}C, ^{15}N]-leucine and the forearm model." *Eur. J. Clin. Invest.* 15 (1985): 349-354.

el Khory, A.E., M. Sanchez, N.K. Fukagawa, V.R. Young. "Whole body protein synthesis in healthy adult humans." *Am. J. Physiol.* 268 (1985): E174-E184.

France, J., C.C. Calvert, R.L. Baldwin, K.C. Klasing. "On the application of compartmental models to radioactive tracer kinetic studies of in vivo protein turnover in animals." *J. Theor. Biol.* 133 (1988): 447-471.

Hsu, H., Y.M. Yu, J.W. Babich, J.F. Burke, E. Livni, R.G. Tompkins, V.R. Young, N.M. Alpert, A.J. Fischman. "Measurement of muscle protein synthesis by positron emission tomography with L-[methyl-11C]-methionine." *Proc. Natl. Acad. Sci. USA* 93 (1996): 1841-1846.

Krawielitzki, K., T. Volker, S. Smulikowska, H. Bock, and J. Wunsche. "Further studies on the multi-compartmental model of protein metabolism." *Arch. Tierernahr.* 27 (1977): 609-621.

Nissim, I., M. Yudkoff, S. Segal. "A model for determination of total body protein synthesis based on compartmental analysis of the plasma [^{15}N] glycine decay curve." *Metabolism* 32 (1983): 646-653.

Samarel, A.M. "In vivo measurements of protein turnover during muscle growth and atrophy." *FASEB J.* 5 (1991): 2020-2028.

Wolfe, R.R. "Empirical assessment of model validity." *JPEN-J. Parenter. Enteral Nutr.* 15 (1991): 50S-54S.

Lipoproteins, Triglyceride (triacylglycerol), and Fatty Acids

Foster, D.M., P.W. Barrett, G. Toffolo, W.F. Beltz, C. Cobelli. "Estimating the fractional synthetic rate of plasma apolipoproteins and lipids from stable isotope data." *J. Lipid Res.* 34 (1993): 2193-2205.

Harris, W.S., W.E. Connor, D.R. Illingworth, D.W. Rothrock, D.M. Foster. "Effects of fish oil on VLDL triglyceride kinetics in humans." *J. Lipid Res.* 31 (1990): 1549-58.

Hultin, M., R. Savonen, T. Olivecrona. "Chylomicron metabolism in rats: lipolysis, recirculation of triglyceride-derived fatty acids in plasma FFA, and fate of core lipids as analyzed by compartmental modeling." *J. Lipid Res.* 37 (1996): 1022-1036.

Phair, R.D., M.G. Hammond, J.A. Bowden, M. Fried, W.R. Fisher, M. Berman. "Preliminary model for human lipoprotein metabolism in hyperlipoproteinemia." *Fed. Proc.* 34 (1975): 2263–2270.

Sauerwald, T.U., D.L. Hachey, C.L. Jensen, H. Chen, R.E. Anderson, and W.C. Heird. "Effect of dietary alpha-linolenic acid intake on incorporation of docosahexaenoic and arachidonic acids into plasma phospholipids of term infants." *Lipids* 31 (Suppl. 1996): S131–S135.

Zech, L.A., R.C. Boston, D.M. Foster. "The methodology of compartmental modeling as applied to the investigation of lipoprotein metabolism." *Meth. Enzymol.* 129 (1986): 366–84.

Zech, L.A., S.M. Grundy, D. Steinberg, M. Berman. "Kinetic model for production and metabolism of very low density lipoprotein triglycerides." *J. Clin. Invest.* 63 (1982): 1262–73.

Zech, L.A., E.J. Schaefer, T.J. Bronzert, R.L. Aamodt, and H.B. Brewer, Jr. "Metabolism of human apolipoproteins A-I and A-II: a compartmental model." *J. Lipid Res.* 24 (1983): 60–71.

Insulin and Glucose

Cobelli, C., and G. Pacini. "Insulin secretion and hepatic extraction in humans by minimal modeling of C-peptide and insulin kinetics." *Diabetes* 37 (1988): 223–231.

McGuire, E.A., J.D. Tobin, M. Berman, R. Andres. "Kinetics of native insulin in diabetic, obese, and aged men." *Diabetes* 28 (1979): 110–120.

Pacini, G. "Mathematical models of insulin secretion in physiological and clinical investigations." *Comp. Meth. Prog. Biomed.* 41 (1994): 269–285.

Ward, G.M., D.M. Weber, I.M. Walters, P.M. Aitken, B. Lee, J.D. Best, R.C. Boston, and F.P. Alford. "A modified minimal model analysis of insulin sensitivity and glucose-mediated glucose disposal in insulin-dependent diabetes." *Metabolism* 40 (1991): 4–9.

Ketone Bodies

Hall, S.E., M.E. Wastney, T.M. Bolton, J.T. Braaten, and M. Berman. "Ketone body kinetics in humans the effects of insulin-dependent diabetes, obesity, and starvation." *J. Lipid Res.* 25 (1984): 1184–1194.

Wastney, M.E., S.E. Hall, M. Berman. "Ketone body kinetics in humans: a mathematical model." *J. Lipid Res.* 25 (1984): 160–174.

Physiological and Pharmacokinetic Models

Balant, L.P., and M. Gex-Fabry. "Physiological pharmacokinetic modelling." *Xenobiotica* 20 (1990): 1241–1257.

Kohn, M.C., G.W. Lucier, G.C. Clark, C. Sewall, A.M. Tritshcer, and C.J. Portier. "A mechanistic model of effects of dioxin on gene expression in the rat liver." *Toxicol. Appl. Pharmacol.* 12 (1993): 138–154.

Leung, H.-W., D.J. Paustenbach, F.J. Murray, and M.E. Andersen. "A physiological pharmacokinetic description of the tissue distribution and enzyme-inducing properties of 2,3,7,8-tetrachlorodibenzo-p-dioxin in the rat." *Toxicol. Appl. Pharmacol.* 103 1990): 399–410.

Luecke, R.H., W.D. Wosilait, B.A. Pearce, and J.F. Young. "A physiologically based pharmacokinetic computer model for human pregnancy." *Teratology* 49 (1994): 90–103.

Portier, C., A. Tritscher, M. Kohn, C. Sewall, G. Clark, L. Edler, D. Hoel, G. Lucier. "Ligand/receptor binding for 2,3,7,8-TCDD: Implications for risk assessment." *Fund. Appl. Toxicol.* 20 (1993): 48–56.

β-carotene

Novotny, J.A., S.R. Dueker, L.A. Zech, A.J. Clifford. "Compartmental analysis of the dynamics of beta-carotene metabolism in an adult volunteer." *J. Lipid Res.* 36 (1986): 1825–38.

Vitamin A

Green, M.H., J.B. Green, T. Berg, K.R. Norum, R. Blomhoff. "Vitamin A metabolism in rat liver: a kinetic model." *Am. J. Physiol.* 264 (1993): G509–G521.

Vitamin B6

Coburn, S.P., and D.W. Townsend. "Modelling vitamin B6 metabolism in rodents." *In Vivo* 3 (1989): 215–223.

Zempleni, J. "Pharmacokinetics of vitamin B6 supplements in humans." *J. Am. Coll. Nutr.* 14 (1995): 579–586.

Vitamin D

Leeuwenkamp, O.R., H.E. van der Wiel, P. Lips, W.J. van der Vijgh, R. Barto, H. Greuter, J.C. Netelenbos. "Human pharmacokinetics of orally administered (4R)-hydroxycalcindiol." *Eur. J. Clin. Chem. Clin. Biochem.* 31 (1993): 419–426.

Vitamin E

Traber, M.G., R. Ramakrishnan, and H.J. Kayden. "Human plasma vitamin E kinetics demonstrates rapid recycling of plasma RRR-alpha-tocopherol." *Proc. Natl. Acad. Sci. USA* 91 (1994): 10005–10008.

Calcium

Staub, J.F., P. Tracqui, S. Lausson, G. Milhaud, A.M. Perault-Staub. "A physiological view of in vivo calcium dynamics: the regulation of a non-linear, self-organized system." *Bone* 10 (1989): 77–86.

Iron

Gupta, M.M., P. Roth, and E. Werner. "Changes of erythropoietic and storage iron components in certain clinical situations as evaluated by ferrokinetic investigations." *Clin. Phys. Physiol. Meas.* 13 (1992): 411–418.

McLaren, G.D., M.H. Nathanson, A. Jacobs, D. Trevett, and W. Thomson. "Regulation of intestinal iron absorption and mucosal iron kinetics in hereditary hemochromatosis." *J. Lab. Clin. Med.* 117 (1991): 390–401.

Zinc

Scott, K.C., and J.R. Turnlaund. "A compartmental model of zinc metabolism in adult men used to study effects of three levels of dietary copper." *Am. J. Physiol.* 267 (1994): E165–E173.

Wastney, M.E., R.L. Aamodt, W.F. Rumble, and R.I. Henkin. "Kinetic analysis of zinc metabolism and its regulation in normal humans." *Am. J. Physiol.* 251 (1986): R398–R408.

Wastney, M.E., I.G. Gokmen, R.L. Aamodt, W.F. Rumble, G.E. Gordon, and R.I. Henkin. "Kinetic analysis of zinc metabolism in humans after simultaneous administration of ^{65}Zn and ^{70}Zn." *Am. J. Physiol.* 260 (1991): R134–R141.

Heavy Metals

De Michele, S.J. "Nutrition of lead." *Comp. Biochem. Physiol. A* 78 (1984): 401–408.

Urea, Ammonia, and Creatinine

Fujii, T., M. Kohno, and C. Hiriyama. "Metabolism of ^{15}N-ammonia in patients with cirrhosis: a three-compartmental analysis." *Hepatology* 16 (1992): 347–352.

Kay, J.D., J.W. Seakins, D. Geiseler, and M. Hjelm. "Validation of a method for measuring the short-term rate of urea synthesis after an amino acid load." *Clin. Sci.* 70 (1986): 31–38.

Odeh, Y.K., Z. Wang, T. Ruo, T. Wang, M.C. Frederiksen, P.A. Pospisil, and A.J. Atkinson, Jr. "Simultaneous analysis of insulin and ^{15}N$_2$-urea kinetics in humans." *Clin. Pharmacol. Ther.* 53 (1993): 419–425.

Thomaseth, K., and M. Ingeno. "Simulation of urea kinetics and fluid balance during hemodialysis." In: *Simulation in the Medical Sciences*, Anderson, J. G. and M. Katzper, eds. San Diego: Society for Computer Simulation; 1997: 154–159.

Index

WINDOWS® and MACINTOSH® Version 0-387-94996-8